RS•C

Using ICT to enhance teaching and learning in chemistry

Compiled and written by Steve Lewis
RSC Teacher Fellow 2002–2003

Written by Steve Lewis

Edited by Colin Osborne and Maria Pack

Designed by Imogen Bertin

Published and distributed by Royal Society of Chemistry

Printed by Royal Society of Chemistry

For further information on other educational activities undertaken by the Royal Society of Chemistry write to:

Education Department
Royal Society of Chemistry
Burlington house
Piccadilly
London W1J 0BA

Information on other Royal Society of Chemistry activities can be found on its websites:
http://www.rsc.org
http://www.chemsoc.org
http://www.chemsoc.org/LearnNet contains resources for teachers and students from around the world.

ISBN 0–85404–383–7

British Library Cataloguing in Publication Data.

A catalogue for this book is available from the British Library.

Contents

RS•C

RS•C

Foreword

Information and Communication Technology (ICT) has revolutionised the ways we process, access and use information and, as computers and other devices become ever more powerful and information becomes more readily available, the next generation will need to be able to interact with digital media effectively to exploit these amazing new technologies to their full potential for the benefit of society.

It is therefore imperative that teachers become familiar with ICT and its true potential and can present information with a perspective similar to that which the present generation of young people is using to develop their interests in their everyday life.

This resource from the RSC gives teachers of chemistry practical help needed to integrate ICT into their teaching and stimulate the enthusiasm of a new generation of scientists in the exciting new areas of chemistry that are opening up such as N&N (Nanoscience and Nanotechnology). Furthermore it will be highly effective in developing the new ethos of sustainability that will be a major driving force behind the next advances in chemistry that are vital if we are to survive the manifold problems confronting society in the next few decades.

Professor Sir Harry Kroto BSc PhD CChem
Hon FRSC FRSE FRS

President, Royal Society of Chemistry

Introduction

Whilst I was working on this project the fifth instalment of the Harry Potter saga was published to the accompaniment of extraordinary levels of interest around the world. Part of the appeal of these books is the magical classrooms where, at the wave of a wand, writing appears by itself, illusions are displayed in thin air and objects materialise, are transformed or vanish.

I was struck by how much of this can already happen in a chemistry laboratory without resort to magic or even much technology. I was also struck by how computer technology is an opportunity to add the automatic writing and illusions to make the wonder of chemistry apparent to even the least imaginative of students. Images and animations can be displayed to illustrate the arm waving indulged in by most teachers and help to illuminate what are often perceived as obscure and difficult ideas.

This project only scratches the surface of what is possible using ICT within chemistry teaching. The focus of these guidelines is to suggest some ways in which hardware and software resources already available to many chemistry teachers might be used to create materials that suit the particular needs of individual teachers. It assumes no more technical skill than that currently possessed by most chemistry teachers. Making use of these guidelines and examples of materials may allow chemistry teachers to further develop their expertise in this area and consequently have more influence on future developments.

These guidelines and the examples of materials are a digest of ways in which some chemistry teachers across the UK currently use ICT to develop their own resources. Outline instructions for creating the examples are included, and the publication is structured to be dipped into selectively.

A characteristic of ICT is that it is continually changing. This means that the design of some of the project materials will inevitably become outmoded as the software used to create them evolves. However, even if the skills used to create activities may be short term ones, the ideas themselves are not. It is hoped that the examples included here will prove to be sufficiently useful that they will be updated and improved in step with the development of software and teacher expertise.

As I gathered ideas and materials from colleagues around the country during the project year it was very striking that teachers working in isolation have arrived at broadly similar ways of using ICT. Many of the features and ideas illustrated and discussed here have been developed several times independently. At the same time, individual teachers bring their own insights and techniques which the wider community would benefit from sharing. ICT can make geographical isolation irrelevant and allow interested teachers an efficient opportunity to share and develop their ideas and materials. If they wish, users of these guidelines may become involved in this process through submitting what they create or adapt to a shared resource bank hosted at the project website, **http://www.ChemIT.co.uk**.

Steve Lewis, September 2003

RS•C

Acknowledgements

I am very appreciative of the many comments and encouragement from people who have downloaded draft materials from the ChemIT website and from the participants at several teacher professional development events during the year. Much of this advice has been incorporated within these guidelines, so that they genuinely reflect ways in which chemistry teachers are using ICT with their students.

Colin Osborne, Maria Pack, Bonnie Howard and Ted Lister of the Royal Society of Chemistry have been a constant and patient source of support throughout the project year.

Special credit and thanks for frequent advice and in many cases the donation of materials to the project, are due to the following colleagues:

Vicky Askew	Manchester Grammar School
Warwick Bailey	University of Cambridge (CARET) for examples of Java Applets **http://www.scormtech.com/ chem/applets/index.htm**
Rachel Biggs	King Edward VI Stratford-upon-Avon
Elaine Blenkharn	Ludlow College
Ian Bridgewood	Queen Elizabeth II Sixth Form College
Judith Brown	Barnsley College
Alan Carter	Wellington School
Lawson Cockcroft	Consultant
Derek Denby	John Leggott Sixth Form College
David Davies	British School in the Netherlands
Neil Dixon	South Bromsgrove High School
Steve Dixon	Haileybury School
Mike Docker	Farnborough Sixth Form College **http://www.mp-docker.demon.co.uk/**
Chris Evans	Queen Elizabeth's Grammar School, Blackburn **http://www.btinternet.com/ ~chemistry.diagrams/**
Roger Frost	ICT consultant **http://www.rogerfrost.com/**
Theodore Gray	Author **http://www.theodoregray.com/ PeriodicTable/**
Frank Harris	Malvern College
Karl Harrison	Oxford University for examples of Shockwave animations **http://www.chem.ox.ac.uk/it/**
Peter Hollamby	St Cyres School, Penarth
Joan Hosfield	Cadbury Sixth Form College
Ewen McLaughlan	Swansea College
Christine Milsom	Greenhead Sixth Form College
Andrew Page	Braintree College
Emma Rees	West Somerset Community College
Malcolm Stephen	Mackie Academy, Stonehaven
Dave Tandy	Solihull Sixth Form College
Mike Thompson	Truro School **http://www.chem-pics.co.uk**
Keith Wilkinson	The International School, Lusaka, Zambia **http://members.lycos.co.uk/ chemistry/**
Antony Williams	ACD / ChemSketch, Toronto **http://www.acdlabs.com**

The Royal Society of Chemistry would like to extend its gratitude to the Principal and governors of Shrewsbury Sixth Form College for seconding Steve Lewis to the RSC's Education Department.

Using ICT in chemistry teaching:
an overview

- ICT can open a window from your classroom into other worlds: molecular and macroscopic, real and imaginary, past and future. You can use it to reach further down the corridor or to the other side of the planet.

- Chemistry specifications can often seem remote or irrelevant to real life - ICT can provide opportunities to make connections explicit or to visualise previously unexpected ones.

- Much of chemistry thinking involves constructing and contemplating models for the behaviour of particles at a molecular level or their macroscopic effects - ICT can provide numerous ways of visualising and explaining these.

- ICT can connect teachers, pupils and practising chemists however isolated they may be geographically.

- Regular use of ICT with students can lead to it becoming second nature, so that you are able to concentrate on teaching and learning rather than worrying about technical issues.

- ICT is in a state of continual change: the format of ICT materials means they can be accessed, customised, updated and exchanged far more easily than paper based materials.

- Only use ICT if it enriches what you do already.

- ICT shouldn't replace existing good practice - but it can enhance it.

- Be aware that ICT based learning isn't necessarily the ideal for all students, but one advantage it has is that it can allow all students to target their individual needs: ICT can provide students with more choices about how they are going to learn.

- Don't expect to be able to purchase 'off the shelf' solutions to address a need to include ICT within your teaching – if you do find any, treasure them!

- Evaluate the ICT facilities available to you now and requiring your current level of skill, make regular use of these, develop from there and in the process become more discerning about the quality of commercial products.

- Students need to be to able to provide hand written responses to questions on paper – ICT rarely provides opportunities to practice this.

- Chemistry teachers are not IT technicians and it is unreasonable to expect them to behave as such, consequently fostering good relations with technical support staff can be time well spent.

- Electronic simulations and video clips are no substitute for carrying out practical work, however they can allow more benefit to be derived from practical experiences.

- Beware of superficially impressive presentation in ICT materials – always question how effectively a resource meets each individual student's needs.

- To make a start with these materials: choose one activity, try it, evaluate it rather than attempting too much at once.

- Concentrate on developing ICT materials that target specific learning needs, quality rather than quantity.

- Students say that they place a high value on being able to use ICT based materials in their own time that have previously been used with a teacher in the classroom.

- Bids for time and / or resources to develop the use of ICT within a department may be strengthened by referring to the materials within this project as concrete examples of what can be done.

- Chemistry teachers say they prefer small units of ICT based material that they can integrate into their own teaching programmes - as opposed to resources that essentially mean delivering another teacher's lesson.

- Be aware that there are a range of technologies available to display and interact with output from a computer in the classroom and that the method chosen can affect teaching and learning experiences – interactive whiteboards are only one of several methods.

RS•C

The CDROM and ChemIT website

The CDROM accompanying these guidelines contains a range of examples of ICT activities, including many contributions from colleagues. It is not intended to be a comprehensive resource bank, but rather to provide materials to be copied and edited.

The same materials are available for download from the project website at **http://www.ChemIT.co.uk** and it is intended that the materials here may be supplemented with contributions from colleagues who have made use of the project guidelines to develop their own materials. In this way new activities may be shared widely between chemistry teachers.

The examples referred to within these guidelines have a unique reference number which can be used to find them both on the CDROM and the ChemIT website.

Resources include the following:

- Microsoft Office® files (PowerPoint®, Word® and Excel®) which can be used as provided, edited or as templates for new materials;

- Samples of Java Applets, Flash animations, animated gifs and images.

Copyright statement

The electronic resources and content within them are copyright free for non profit educational work, as long as the original source is clearly referenced. They may be networked or used on a standalone computer.

Technical problems

Drafts of these activities were widely tested during the development year. The RSC can make no guarantee that they will work on every computer system and cannot offer any technical support.

Sharing resources

It is intended that the ChemIT website should be a means of sharing new ICT resources developed by teachers of chemistry, including those created through applying these guidelines. In this way the resources can keep pace with changes in software and include new ideas for using ICT with students. It can also become more comprehensive, covering a broader range of the 11–19 chemistry curriculum. This provides an opportunity for any teacher of chemistry to take part in the project and shape it's development and reflects the fact that ICT is an aspect of teaching and learning undergoing continuous evolution.

Any contributions will be screened by the RSC in terms of their chemical content and so as not to duplicate what is already available on the site.

See the ChemIT website for more details
http://www.ChemIT.co.uk and
http://www.chemsoc.org/learnnet/chemIT

Using these guidelines

Inevitably teachers looking through these guidelines will have a range of interests and experiences related to making use of ICT within the teaching of chemistry. What they are all likely to have in common is limited time.

To support this diversity the guidelines have been written so that they may be used in various ways, for instance:

- they can be read from beginning to end to build a coherent view of possibilities and how these might be achieved;

- an individual section can be read in isolation;

- a particular resource suggested in **Integrating ICT into a sequence of chemistry lessons** can be created by referring to the relevant section in the Skills guidelines.

This is achieved at the cost of some repetition of key points. The following overview provides a brief description of the contents of each section to support quick browsing.

The guidelines refer to a range materials produced for the project. These are intended to provide examples of the different ideas rather than comprehensive coverage of the 11–19 chemistry specifications. These materials are available on the accompanying CDROM and both from the project website (**http://www.ChemIT.co.uk**) and via the RSC learnnet website (**http://www.chemsoc.org/learnnet/ChemIT**).

Each individual resource has been given a unique reference number (quoted within these guidelines) to make it as easy to find.

Many of the section headers are questions that have been raised by chemistry teachers during training sessions exploring how ICT might be used to enrich their teaching.

A note about weblinks (URLs) in the text

The addresses of several websites are included within the guidelines. The ChemIT website also contains these links so that any changes can be updated and also so that any link may be followed from the site using 'one click', rather than by laboriously typing the address into an internet browser.

Integrating ICT into a sequence of chemistry lessons

Why use ICT?

Comments from a Chemistry teacher provided with ICT facilities in her classroom:

'At first I resented using the ICT kit in my lessons as I thought I could do much the same much more quickly without the technology. However, since I've overcome the "activation energy" of getting used to the computer I've changed my mind. I find I can make my lessons much more stimulating and often forget that I'm even using ICT to do so!'

Some factors that might encourage chemistry teachers to strive to overcome the 'activation energy' referred to above are outlined below. The views expressed here are based on comments from a range of learners and teachers. The resource reference numbers refer to materials developed for the project which illustrate some of the points made.

Clarity

Students often comment with approval on relatively mundane aspects of materials prepared using ICT. When information is displayed on a screen learners say features such as typed text, colour for emphasis and an increase in pace can all be helpful, *eg* when it is necessary to focus on a particular piece of a more complex picture where there may be a lot of distracting information. This might be particularly relevant to chemists where chemical symbolism and language can make it difficult for students to 'filter out' key points.

It has been reported that making use of 'electronic' displays like this can lead to marked improvements in the quality of notes taken by students.

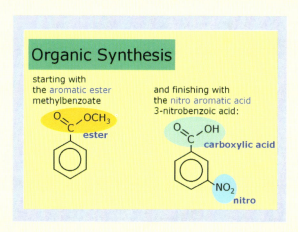

Figure 1
Slide from a PowerPoint® presentation (resource 1) supporting an organic preparation. The text and images appear in a pre-set sequence.

RS•C

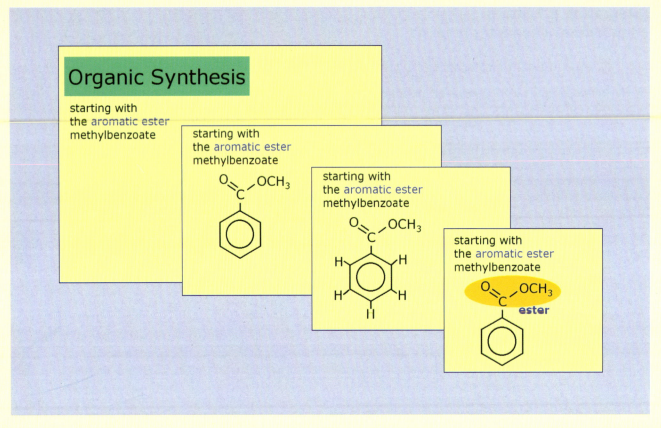

Figure 2

Steps from the animation sequence from the Powerpoint® presentation (resource 1) supporting an organic preparation leading to the slide illustrated in Figure 1. Students can be engaged by prompts to suggest 'what happens next' before the next image is revealed.

The ability to reveal a sequence of ideas or a diagram in stages and replay this process supports the 'constructive' nature of much chemistry teaching, where compound ideas are built up as logical sequences.

A video camera or microscope connected to a computer can make a practical demonstration more visible to the whole class. The camera can monitor a view in 'real time' without recording or can record for future use, eg as a reminder of what happened.

Inspiration

ICT has the potential to provide a window to anywhere in the universe: microscopic or macroscopic, everyday or imagined. Examples include being able to tell chemical stories illustrated with still images or video clips or to display an interesting website.

Flexibility

Using ICT means being able to respond to the needs of a class and change the direction of a lesson very quickly. For example, to review the background to an idea that has proved unexpectedly difficult, to stimulate a sleepy group with a quick activity such as a quiz (resource 3) or context story (resources 4 and 5), to have resources on hand to support a 'lets revise anything you like' session.

Adaptability

Using generic software such as Microsoft Office® applications allows the design of electronic materials that are easy to update, for instance, to reflect changes in curriculum specification, experience in using them, a different approach to teaching or different demands from learners. This is in contrast to traditional resources such as 'cut and paste' handouts or overhead transparencies where any changes mean that the resource needs to be made all over again.

For instance, the project materials include animated PowerPoint® presentations of organic reaction mechanisms (resources 6, 7 and 8). These are currently in a form that is probably most useful at the end of a post-16 course. However, it is straightforward to copy the slides for a particular reaction and mechanism and save these as part of a new activity, perhaps to introduce or review a discrete topic in organic chemistry or to support a laboratory preparation.

Figure 3
An image displayed using a Powerpoint® slide (resource 2). These can be related to a chemical context.
Photo: J Bump

Interactivity

Materials can be designed to encourage and guide thinking about problems. Questions or prompts to think can appear before a model answer, so that a learner is guided through a particular thought process.

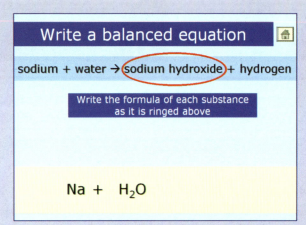

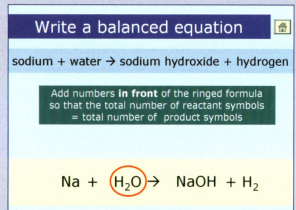

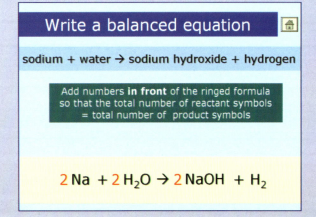

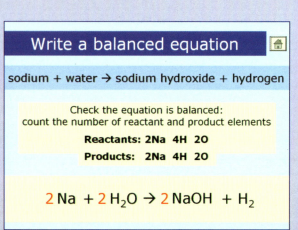

Figure 4

A Powerpoint® slide sequence of detailed prompts to guide a student through the steps to balance a reaction equation. Intended to be used independently (resource 9). The format allows teachers to customise the text to suit their own approach.

Choices can be provided for the learner and in some types
of resource these can provide feedback.

Atomic Structure

- Look at the comments below and decide on the correct answer by clicking on the shaded box.
- Choose the correct response from the drop down menu that appears.
- Then mark your work by looking at the answers on the answer sheet.

First type your name here:

1. Atoms contain **three different particles**. Protons have a (i) *click here* charge, **neutrons** have a (ii) ⬚ charge and **electrons** have a (iii) ⬚ charge.

2. The **nucleus** contains (i) ⬚ and (ii) ⬚.

3. What is meant by the statement: "**All atoms are neutral**"?
   They contain the same number of ⬚.

4. Which of the following **BEST** describes the term **atomic number**?
   It is the number of neutrons in the nucleus. ☐
   It is the number of protons in the nucleus. ☐
   It is the number of electrons in an atom. ☐

5. The **mass number** is the number of ⬚ ▾ the nucleus of an atom.

6. Which of the following **BEST** de⬚ ⬚pes?
   Atoms of the same element tha⬚ | protons and neutrons | mbers of neutrons. ☐
   Atoms of the same element tha⬚ | electrons and neutrons | mbers of electrons. ☐
   Atoms of the same element tha⬚ | protons and electrons | mbers of protons. ☐

7. Which of the following **pairs** are isotopes?

   (i) $^{12}_{6}C$ (ii) $^{15}_{7}N$ (iii) $^{14}_{6}C$ (iv) $^{18}_{8}O$ (v) $^{40}_{19}K$ (vi) $^{40}_{18}Ar$

Figure 5
A Word® document (resource 10) that uses 'drop down' menu choices but no feedback.
Clicking a gap reveals a menu of choices, selecting one of these leads to the text being appearing on the page. Thanks to Peter Hollamby.

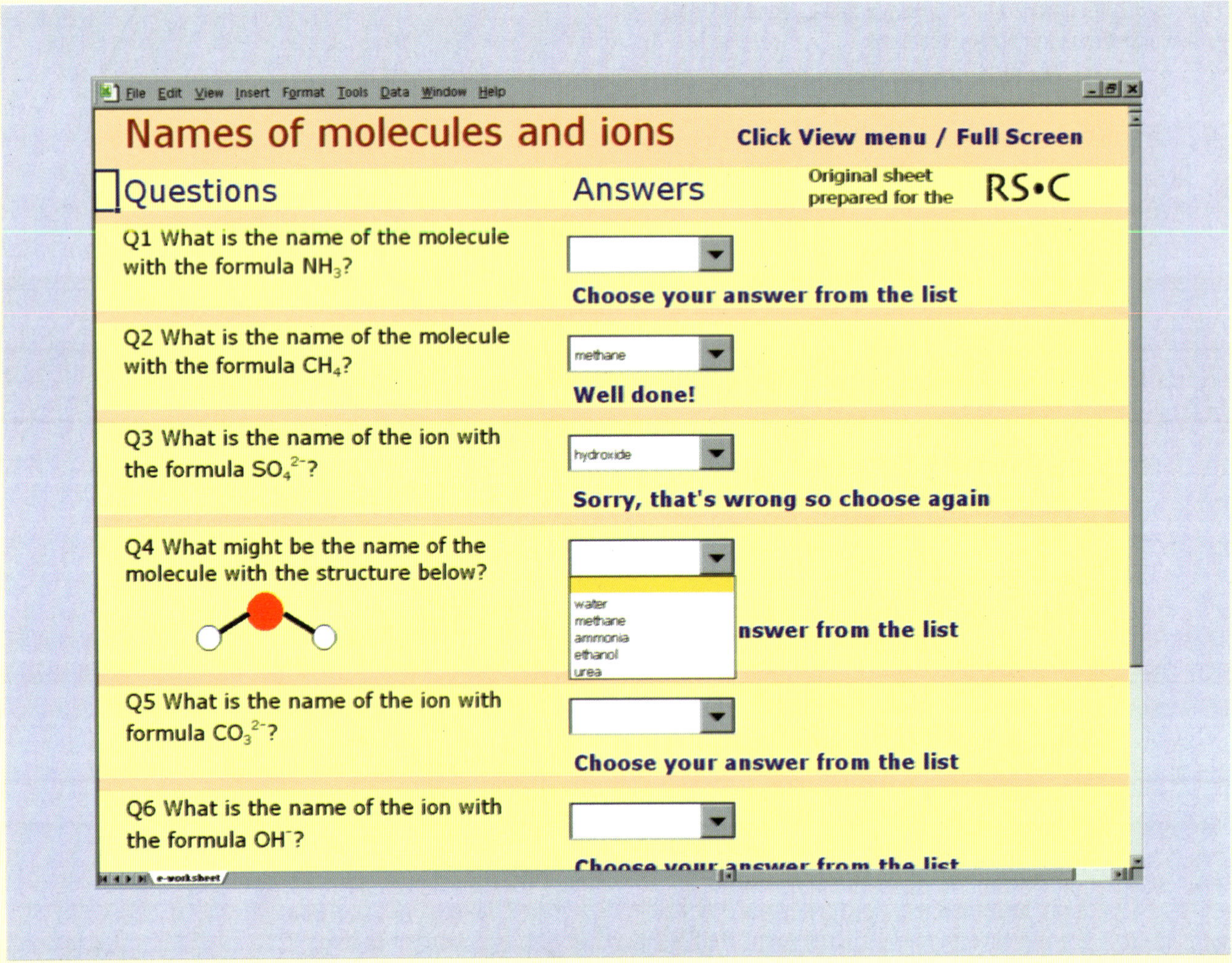

RS•C

File Edit View Insert Format Tools Data Window Help

Names of molecules and ions

Click View menu / Full Screen

| Questions | Answers | Original sheet prepared for the **RS•C** |

Q1 What is the name of the molecule with the formula NH_3?

Choose your answer from the list

Q2 What is the name of the molecule with the formula CH_4?

methane

Well done!

Q3 What is the name of the ion with the formula SO_4^{2-}?

hydroxide

Sorry, that's wrong so choose again

Q4 What might be the name of the molecule with the structure below?

water
methane
ammonia
ethanol
urea

nswer from the list

Q5 What is the name of the ion with formula CO_3^{2-}?

Choose your answer from the list

Q6 What is the name of the ion with the formula OH^-?

Choose your answer from the list

e-worksheet

Figure 6

*An Excel® spreadsheet (resource 11) that uses 'drop down' menu
choices and provides feedback according to the selection made.*

Accessibility

It is possible to provide learners with access to materials in several ways. For example on disk or CDROM, via school or college internal computer networks or external websites, using e-mail attachments or via free internet 'user groups'.

An advantage of having electronic versions of activities available to students is that those who have difficulty reading a printed version can adapt an 'e-version' so that it has larger or clearer font, different colours, or even use a screen reading programme if their eyesight is particularly limited.

Learners frequently comment that they find the opportunity to 'unpack' key parts of a previous lesson in their own time very valuable. They say this is because it enables them to use materials that are familiar to them from that lesson and are able to work through these at their own pace and with their own emphasis.

Feedback repeatedly suggests that using ICT to present and reinforce ideas certainly helps some learners to grasp these ideas more quickly than they would otherwise. This isn't the case to the same extent for all students, but having the opportunity to use ICT with learners and doing so carefully (for instance, not choosing to use it all the time) allows a wider range of learning styles to be supported.

Efficiency and transferability

Large amounts of information and large numbers of different resources may be stored electronically. Materials in this format may be readily altered in response both to experiences when using them and to changes in learner needs. They may be easily and quickly transferred between colleagues and learners.

When used effectively ICT can save time in the classroom and leave more time for other activities, such as more experimental work. For instance the pace of a lesson can be increased: having key points pre-prepared and available to display can save time otherwise spent writing text on a board. Having pre-prepared instructions displayed on a screen to the class can replace paper instructions, saving time in printing and distributing the paper copies.

Managing these opportunities takes some practice. It is a common experience that getting to grips with these can be itself time consuming, but this is offset by the long term benefits.

What constrains the use of ICT?

Some of the factors suggested by various teachers of chemistry that may contribute to the 'activation energy' constraining them from integrating ICT into their teaching are listed below. There are no easy solutions to some of these, though feedback suggests that an increasing number of colleagues are overcoming them.

Lack of time

These guidelines aim to show how it is possible to integrate ICT usefully into all aspects of a chemistry teaching programme. It is unrealistic to expect to have all aspects of this in place within a short period when starting from scratch. However, choosing a clearly defined and limited area of work to start with will lay foundations for further development. For example, aim to create a few animated PowerPoint® slides to support a few difficult ideas within the next half-term's scheme of work, or use some provided in this resource.

Another related issue is the time needed to think of approaches and develop the necessary skills to apply these. These project materials are an attempt to share some of the ideas that chemistry teachers have already developed and pass on some of the skills needed to implement these.

Advice given within these guidelines and techniques used to create the teaching and learning materials have come through the sharing of ideas between chemistry teaching colleagues. These suggestions often relate to things that took a significant amount of time to resolve, yet are very quick and straightforward to describe. Consequently reference to these guidelines can be a significant time saver, may avoid repeating work already undertaken by others and hopefully lead to fewer people giving up in frustration.

Lack of facilities

Having clear ideas of how ICT facilities might be used to enrich teaching and learning makes it more likely that bids for these might be successful.

Lack of technical support

The support and understanding of senior managers are required to ensure that an effective and responsive technical support system is in place. It is unrealistic and unreasonable to expect busy teachers to fulfil this role themselves.

If effective systems have not yet been established in a particular institution then the fact that they have elsewhere might be useful supporting evidence when petitioning for progress, particularly in combination with specific examples of how these facilities could enrich the teaching and learning of chemistry.

Lack of familiarity with the equipment

Worrying about which button to click makes it difficult to focus on teaching and learning. Familiarity comes with practice and even being prepared to make mistakes in front of students.

Lack of personal ICT skills

Unfortunately too many teachers' experience of ICT training is of attending generic skills courses that have little obvious relevance to their classroom needs.

Having a clear idea about which skills are likely to be of direct benefit to teaching and then setting achievable targets for gaining these is a sensible approach. Reference to these guidelines may be of help with this process.

Lack of teaching and learning resources

The materials produced for this project and guidelines for amending or creating them provide a possible starting point from which to build a resource bank tailored to the needs of a particular institution.

When bidding for staff development time it might be useful to refer to these project materials and propose tailoring them to the needs of the institution. Experience suggests that if a clear idea linked to teaching and learning can be proposed then managers are often more receptive when it comes to releasing resources.

Lack of peer support

This can arise for various reasons including geographical isolation or colleagues' exhaustion, cynicism, negativity or inertia. The resources for this project represent a collection of ideas from the teaching community as a whole and therefore provide an opportunity to share these.

The 'C' in ICT is for communication and the technology allows teachers of chemistry to become involved in 'electronic staff rooms' where electronic mail can be used to discuss problems and ideas and rapidly exchange electronic resources between individuals who are separated geographically.

Lack of motivation

Colleagues who have already experimented with these approaches repeatedly comment enthusiastically on the many benefits that have arisen from their having made an effort to develop skills in order to be able to integrate ICT resources into their chemistry teaching. Some of these benefits are dealt with above, but the key points seem to be improved student motivation (and in some cases attainment), improved flexibility in teaching and more enjoyment of the subject from both the learner and teacher.

How might a computer display be projected to a large screen?

An ever increasing proportion of chemistry teachers are able to display information from a computer on a large screen within their classroom. Feedback suggests that learners enjoy interacting with materials in this way and some may even expect it. They particularly appreciate having access to ICT materials in their own time that they have first met in class. Making clear connections between materials introduced in the classroom and independent activities carried out elsewhere appears to be a key factor in enabling full and effective use of these.

However, if teachers do not have the facilities to display information in this way the types of ICT resource used can still be valuable to learners when presented in other ways, for example via the school or college computer network or website, or on disk for use at home.

There are various ways of projecting computer output and of controlling this. Interactive whiteboards are not the only method available to teachers, although the impression is often given that they are. The method chosen can have a strong influence on teaching styles and learner experiences.

For instance, a teacher who generally uses a blackboard creates a different learning environment from one using an overhead projector (OHP). Neither approach is the correct one yet teachers have their own preferences and become comfortable with these. Teachers preferring to work directly at a board may be happiest using electronic whiteboards, where they interact directly on the screen displaying information whilst turning their backs on their class. In contrast, using a mouse and keyboard, graphics tablet or 'tablet' computer to control what appears on the screen is more akin to using an OHP, where the teacher is distant from the information displayed and can face the group at all times.

A less expensive alternative for displaying to a class is to use one or more big screen television monitors connected to the computer, though this does not provide as large a screen area as a data projector.

The table opposite summarises some of these options.

Projection method	Control of display	Points to consider
Data projector	Electronic whiteboard	Teacher interacts directly with the image displayed, has back to group whilst doing so and can cast a shadow on the screen. Instant hand writing on the screen. Generally fixed in place and those that are not are awkward to move.
Data projector	Graphics tablet	Teacher is remote from the image and can face a group at all times. Allows instant hand writing displayed on the screen. Some graphics tablets can be used anywhere in the classroom. Easily connected to any computer and projection system.
Data projector	Mouse and keyboard	Teacher remote from image and can face group at all times. Cannot produce instant hand writing on the screen. A 'wireless' mouse is a relatively inexpensive way of allowing the content on the screen to be controlled from various places within the classroom.
Data projector	Touch sensitive monitor eg with a 'tablet' PC	Similar functionality as graphics tablet but more expensive.
Single or grouped television monitors connected to the computer	All of above	Cheaper than using a data projector but a smaller display area. Using several screens around the room divides the focus of attention for a class.

An example of ICT facilities within a chemistry laboratory at Solihull Sixth Form College. With this arrangement, teachers of chemistry can face the class and use an unobtrusive computer monitor to view what is projected on the screen behind them. Note that the screen here is not an interactive whiteboard.

Ceiling mounted data projector

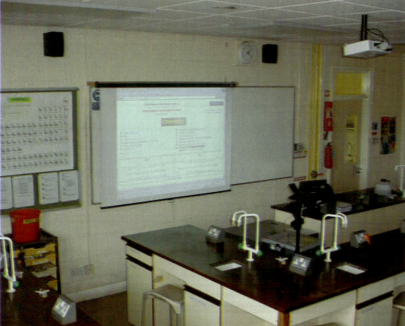

Computer monitor on the teacher's desk with mouse and keyboard

Computer and video player stored beneath the teacher's desk

Comments from a chemistry teacher who has been experimenting for some time with making use of ICT with his students:

'I would like to see short, snappy resources that I can integrate easily into my teaching, something that does not control the direction of my lesson. I want resources that enable students to visualise chemistry in novel ways. I don't want a set of notes on a web-page unless it is effectively integrated with animation/video and instant feedback questions.'

This section includes many ideas for making use of technology within chemistry teaching. They are a digest of what many teachers of chemistry are already doing and thinking with respect to using ICT to enrich their students' teaching and learning experiences.

A sensible approach, particularly if relatively inexperienced in this area, would be to take one idea, use it and assess its effectiveness rather than risk being overwhelmed through attempting too much at once.

Someone skilled in using ICT to create resources for teaching is not automatically skilled in teaching using ICT. It is a common experience that great resources don't necessarily mean great lessons and, as ever, it's often not what you use but way that you use it that is important.

There is no suggestion here that ICT based activities should displace established successful classroom activities. Paper based and experimental activities are often by far the most appropriate things to be doing in the classroom.

ICT can support many long established and proven procedures and make them even more effective. It can bring exciting new experiences into the classroom. However it is important to continually check that its use is providing a more effective means of doing something.

Although it is common sense not to let technology drive teaching, it inevitably does have an impact on teaching style and this needs to be continually questioned and assessed. This process of reflection can only be a good thing, and hopefully can lead to better teaching, both in relation to using the technology and not. Whilst the use of ICT materials in themselves may not necessarily make for better teachers, reflection on how and when it is appropriate to use these may have this effect.

Using ICT will lead teachers to question how their students learn. An obvious answer is 'in lots of different ways' and ICT can provide effective means to support varied needs.

'Electronic' lessons

Being able to display appropriate images on a large screen in a chemistry classroom has proved to be a very powerful tool for motivating and informing learners, particularly for those who respond to a strong visual stimulus. Combining text with different types of image in simple animated sequences can make topics clearer and more stimulating. Having pre-prepared presentations can make it easier to control the flow and pace of lessons. Students can be very involved in the process through 'what's coming next' questioning, which may involve them writing possible responses before revealing an answer. Developing this approach in the classroom means that students are more likely to make effective use of this kind of resource when they access them independently.

For many teachers delivering lessons using blackboards, overhead projectors, etc, means painstakingly re-writing content, albeit modified through experiment and experience, year after year. Developing 'electronic' lesson that are saved within an electronic format can make information delivery far more efficient and make more time for other activities. The electronic medium allows links to be made to different materials and resources and enables rapid switching between these according to a lesson plan or in response to changing demands. With time a bank of linked resources can be built up, supporting choice in type and level of materials to use, making lesson delivery more flexible and catering for wider diversities of need.

Materials stored electronically are easily adapted with experience, so the resource can be fluid, ever changing. Teachers working in this way often comment on how they update and amend their materials even as they are using them with students. The confidence to work in this way develops as using technology becomes instinctive, leaving the conscious parts of the mind free to interact with learners.

Practical suggestions are developed below for how electronic resources might be used within the different stages of a generalised sequence of chemistry lessons, together with some indication of how these might support the learning of chemistry.

ICT resources

The discussion here is limited almost entirely to materials that make use of Microsoft® applications and can therefore be edited and adapted by teachers to suit their particular needs.

There is an ever increasing choice of ICT materials available for chemistry teaching and learning and some of these are of high quality. Much is produced using specialist authoring software and often with high production values. This ranges from expensive commercial materials, through subsidised products from industry and educational bodies including the RSC, to materials produced by teachers who make this available free or at low cost, often via the internet. A common feature of these materials is that they cannot be customised and have to be used as provided. Often this amounts to delivering another teacher's lesson.

To make effective use of all the opportunities available, teachers need to develop judgment about which types of ICT-based experience work best with their own learners. It is very useful to develop the skill of being able to extract and apply the best examples from the wide variety of ICT resources available.

For the approaches suggested below the teacher exercises control over the content and appearance of the resources. Trying at least some of these may prove to be a way of developing greater discernment when judging the quality of materials developed by others. This is particularly valuable when evaluating some glossy commercial products where appearances can sometimes be misleading regarding their true educational value.

The comments are a digest of views expressed by teachers contributing to the project or attending training courses related to it during 2002–03 and also of comments from learners of all ages about their experiences of learning through this medium.

The resources discussed below can be interwoven into the natural flow of the lesson. Feedback suggests that emphasising particular parts of a lesson through this approach is more effective than trying to deliver the entire content through an information packed display on the big screen. Used in this way the use of ICT can become second nature, just another tool, and in fact both teacher and learner can become unaware of the technology behind the display they interact with.

The examples below are chosen to illustrate how ICT can be used in each of several distinct stages within a general lesson sequence. The choices of type of ICT resource are arbitrary and each one might be equally valuable used in a different context. This is an illustration of the flexibility possible through developing a collection of relatively small and customisable materials.

The general section, with a wide range of curriculum examples, is followed by a 'case study' where examples that could support a series of energetics lessons are given.

One of the design criteria for these materials is that it should be within the capability of the majority of chemistry teachers to create new materials like these, or customise existing ones. The required level of skill is comparable to what is needed to produce basic word processed documents. Several training events with teachers during 2002–03 have confirmed that this is a reasonable benchmark.

Using presentation software such as PowerPoint® allows various types of resource to be combined to form a coherent whole through use of links that launch other applications as they are required. For instance, it might be relevant to move from PowerPoint® to a spreadsheet activity or a web page and then back.

Taking steps to develop the skills needed to set up and use ICT resources linked in this way, at appropriate points in a lesson and so that the process is largely unconscious, is an appropriate target for professional development for chemistry teachers over the forthcoming years.

Starter materials

As students come into the room an attention grabbing sequence related to how the teacher wants the lesson to start or to immediately focus the students' thoughts is running automatically on the large screen. Possibilities include an animation from a commercial product, a PowerPoint® presentation running automatically or a video clip etc.

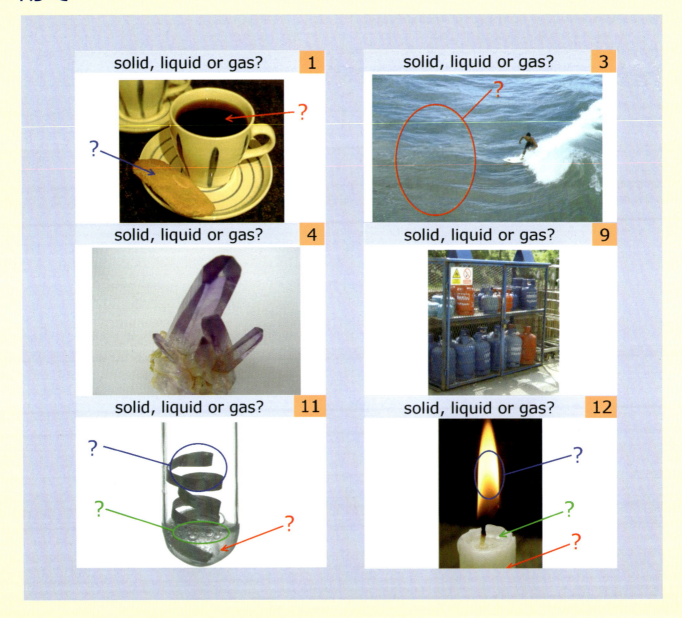

Figure 9

Some of the slides from a PowerPoint file that automatically loops photographs related to solids, liquids and gases (resource 14). This may be used with students of different abilities and levels of experience depending on the question posed by the teacher.

Photos
Coffee cup: Jean Ryder/DHD Multimedia Gallery **http://gallery.hd.org/**
Surfer: James Gray/DHD Multimedia Gallery **http://gallery.hd.org/**
Crystal and gas cylinders: Mike Thompson **http://www.chem-pics.co.uk**
Magnesium plus acid and candle: Peter Hollamby

'Notice board' displays with information (eg this could include deadlines for previous work) for students in one half and an introductory activity in the other might be used on a regular basis. As students become accustomed to these being displayed and the expectation that they respond to them without prompting has been reinforced, teachers are free to attend to start of lesson administration. The notice board file provides a template that can be quickly updated and become familiar to the class.

These 'starter' examples can all operate with minimal involvement from the teacher. However, there are other attention grabbing activities based on many of the materials that follow which a teacher might deliver at the beginning of lesson eg quizzes, context displays using stimulating photographs, video clips or animations. Such activities can enable students to give an indication of an appropriate starting point or level to the teacher.

Figure 10

In the Powerpoint® example below (resource 15), the notice board includes a starter question for the class and a link to a brief answer, which it is important to include if the starter activity is to be closed without encroaching on the main purpose of the lesson.
Spider photo: Damon Hart-Davis / DHD Multimedia Gallery
http://gallery.hd.org/

Deadlines: 05/01/04

Today!
Hand in practical write-up

Weds 24th
Mark worksheet 2

Weds 31st
Topics 4-7 Test

Notices:

Friday 26th
Visit to University labs.
Bus leaves at 8.20
back at 5.30

Activity:
Our blood is **red** due to iron (Fe) ions within haemoglobin.

Damon Hart-Davis / DHD Multimedia Gallery
http://gallery.hd.org/

Spider blood is **blue**.

Suggest a chemical reason for this.

The left side is a full figure illustration.
RS•C

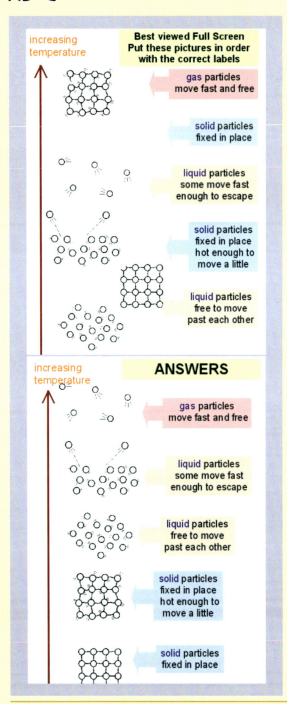

Background

Using ICT facilities can allow efficient and quick review of prior-knowledge before starting a new topic by making use of any ICT material that might be relevant.

Resources produced for a younger age group may be used to review ideas and check misconceptions with older learners. In fact the group may already be familiar with the actual resource used, thus reinforcing the review process. The skill of a teacher in asking appropriate questions can allow the same resource to be used at different levels and for different purposes.

Although developed as an interactive tool for younger students, this may be used at a later stage before moving on to more sophisticated ideas and also to probe understanding.

For instance, discussing the significance of the shaded area between particles in the solid and liquid model illustrations and why this shading is absent in the gaseous ones could be a useful way of exploring a group's concepts of bonding. This could also be used to explore both the strength and weaknesses of the model through consideration of how intermolecular forces are portrayed.

In contrast to a starter activity, exploring the background to a new topic will probably involve a lot of interplay between teacher and learner, with the teacher providing prompts. A presentation could be delivered by one or more students primed in a previous session as part of long term policy where every member of the class uses the ICT facilities to present to the group over several weeks.

Introducing new content

Essential or problematic new points may be introduced through a combination of ICT presentation and teacher demonstration and discourse. The presentation might start with an illustration of a novel application of the new principle, linking to a website with more information that students have the option of following up later. A modelling spreadsheet might be used to demonstrate the effects of changes.

Figure 11
A Word® 'drag and drop' activity (resource 16) that allows descriptions to be matched to images with sample answer.

Emphasising key points and clarifying difficult ideas

Although good teaching involves developing ideas in clear stages in lessons, learners are often left with a complete picture in their notes or text, rather than the logical steps that led to it. Consequently their understanding may be incomplete or confused. A powerful technique that ICT allows, particularly relevant to chemistry teaching, is the opportunity to use presentation software to break down a compound idea into a succession of logical parts and provide an animated record of this process. A very effective feature is being able to instantly go back and re-run parts of a sequence without the need to clear a board and re-write the material.

Many students have commented that they greatly value ways in which ICT can be used to allow them to 're-play' the way an idea or linked body of facts was developed in a previous lesson. They are familiar with the ICT materials and approach since these were introduced in the original lesson and can run through these at their own pace and emphasise whichever parts they choose.

When using presentation software to reveal a linked succession of events (eg as Powerpoint® allows) it is very useful to encourage and develop a 'what happens next?' thinking habit in the learners. Developing this approach in the classroom means that students are more likely to make effective use of the materials when they work through them on their own or in peer groups.

Other Powerpoint® examples include an animated Born-Haber cycle for NaCl (resource 17) and two approaches to the balancing of reaction equations (resource 9).

Feedback suggests strongly that it is not sensible to deliver all of the lesson content in this way, as learners can rapidly become disinterested. What appears to be most effective is using this approach to emphasis key points and, most successfully of all, with concepts that learners find particularly difficult, especially if these involve the combining of several factors.

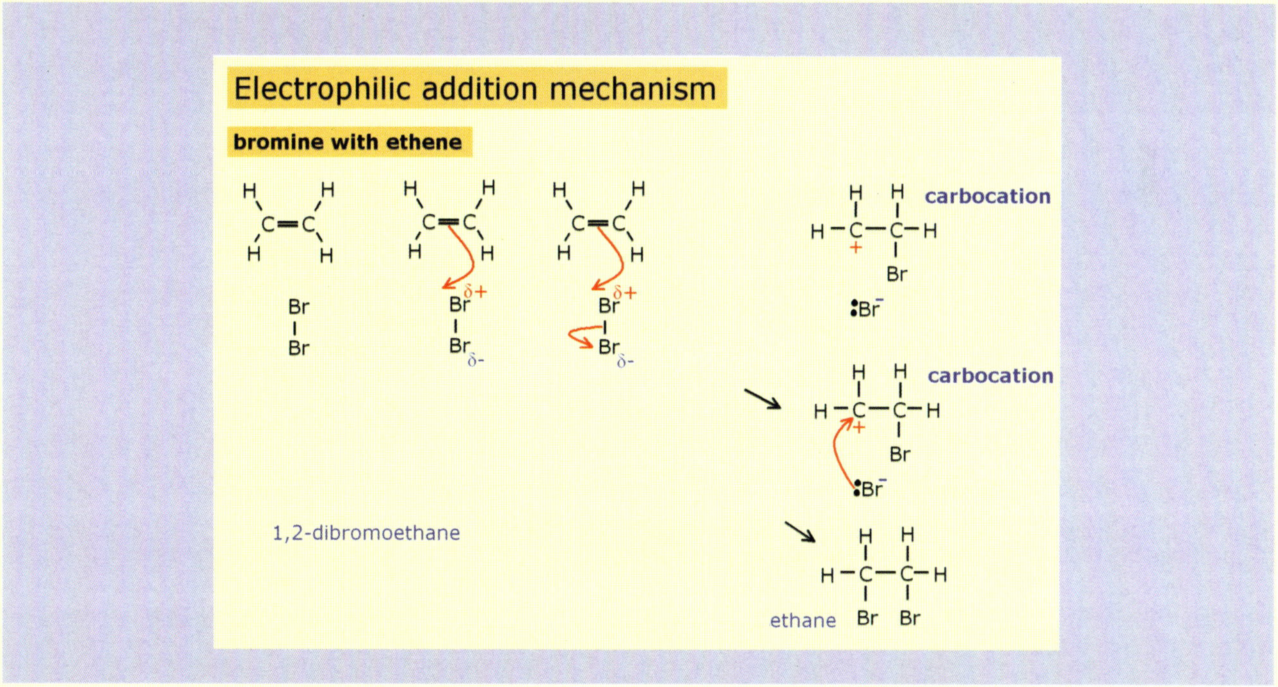

Figure 12
A Powerpoint® organic reaction mechanism (resource 8) sequence revealed step by step by the teacher in class or student independently.

RS•C

Annotation

If it is known in advance that a key feature within a complex image needs to be emphasised then presentation software allows this to be pre-prepared. However, this does not allow for spontaneity and the need to respond to the unexpected during a lesson. If interactive technology is available (interactive whiteboard, graphics tablet etc) then it is possible to use a stylus to write by hand and annotate the screen being displayed.

Powerpoint® does have a 'Pen' option which allows the slide in view to be annotated during a presentation. The annotated screen may not be saved and if the only pointing device available is the mouse, then annotations are of necessity very basic eg coloured lines and highlights rather than text.

Figure 13

A Powerpoint® slide of an infra-red spectrum and annotations to highlight some of its features (resource 1).
Spectrum courtesy of AIST: **http://www.aist.go.jp/RIODB/SDBS/menu-e.html**

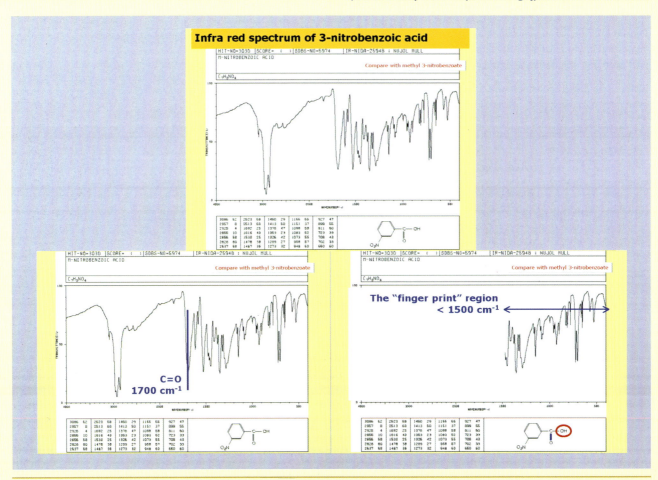

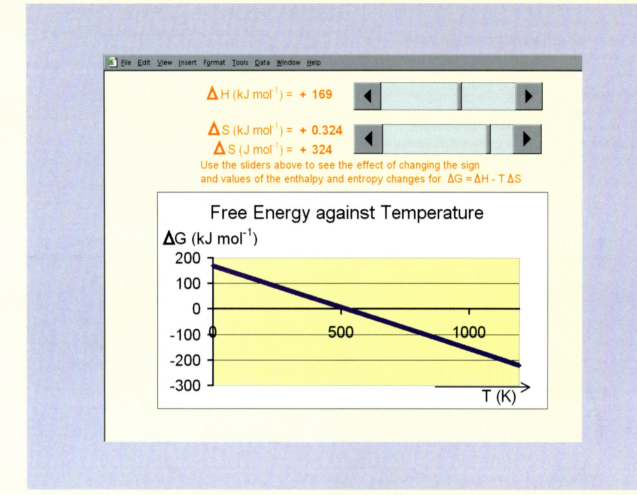

File Edit View Insert Format Tools Data Window Help

ΔH (kJ mol^{-1}) = + 169

ΔS (kJ mol^{-1}) = + 0.324
ΔS (J mol^{-1}) = + 324

Use the sliders above to see the effect of changing the sign
and values of the enthalpy and entropy changes for $\Delta G = \Delta H - T\Delta S$

Free Energy against Temperature

ΔG (kJ mol^{-1})

Modelling

Spreadsheets can be used to create dynamic models that allow the effect of changing variables on a relationship to be investigated.

Context

A powerful feature that arises from being able to display images on a large screen in the classroom is that examples of why chemistry is relevant and important to learners can be presented in a stimulating way. This can be especially useful when covering topics which might be, from the student perspective at least, dry or obscure.

Figure 14

An Excel® spreadsheet (resource 18) that uses 'spinners' to model the effect of changing ΔH and ΔS on a graph of ΔG against temperature.

Having a collection of items readily accessible on the class computer (these might include illustrated anecdotes, everyday chemical applications or things that are simply good fun) means that a teacher has the opportunity to use these with great flexibility. For instance, if a lesson is flagging then introducing one of these for a few minutes can enliven or re-focus the group.

The internet is an excellent source of images that can be used to stimulate interest in chemistry. It is relatively straightforward to find photographs that may be striking, exciting, amusing and unexpected in their relevance to chemistry.

The Skills guidelines cover sources of such images and ways in which these can be included within materials to use with learners. Teachers need to be clear that they have a responsibility to respect any copyright restrictions to the use of images that they may have access to.

Developing understanding – supporting activities

The session may move on to a related discussion, paper-based or practical activity that takes up the bulk of the lesson, with learners working in small groups. Pre-prepared solutions to problems might be displayed on the big screen so that learners who have made more rapid progress can check these and move on whilst the teacher supports others.

Figure 15
A Powerpoint® context slide about retinal and linked animations (resource 2).
Photo: Adam Hart-Davis/DHD Photo Gallery **http://gallery.hd.org/**

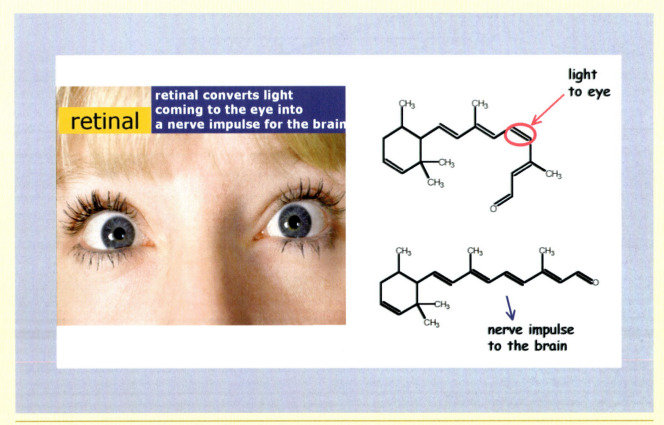

'Drag and drop' activities

A Powerpoint® presentation does not allow objects to be moved around the screen, which can be very useful to use with a class and is a feature of some commercial 'interactive' teaching software packages.

Both Word® and Excel® can both be used to create activities where it is possible to re-organise text and images on the screen by 'dragging and dropping' them using the mouse or other interactive device. The change of state Word® document (resource 16, Figure 11) and the laboratory apparatus activities (resource 12, Figure 7) are examples referred to above.

An advantage of using the generic Microsoft Office® applications over the commercial 'interactive' software to create this type of resource is that the institution has Microsoft Office® software on all of its computers and also students are likely to have the same software at home. Consequently they are able to access these materials without the need to install specialist software.

Students can be asked to do the dragging and dropping and several colleagues have reported that students enjoy interacting with what is displayed to the whole class in this way, particularly those with special educational needs. This approach can provide an alternative way of developing understanding and one that reflects a particular learning style.

Templates

If the classroom has interactive technology in place (eg interactive whiteboard, graphics tablet etc) then pre-prepared templates can be annotated by different students as part of a class discussion.

Examples include apparatus without labels, incomplete reaction equations, incomplete reaction mechanisms, structures without names, names without structures, diagrams without labels, molecules from pictures of their models.

The commercial software provided with products such as interactive whiteboards often allows an annotated screen to be saved so that the work of different students can be compared or accessed at a later date.

Practical work

Large screen presentations are a convenient way of displaying health and safety information (eg Powerpoint® slides of hazard symbols,

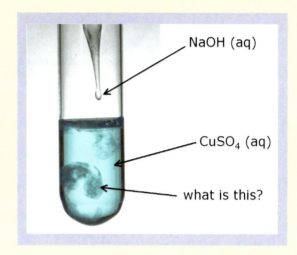

NaOH (aq)

CuSO$_4$ (aq)

what is this?

Figure 16
Powerpoint® slide making use of a photograph of a reaction.
Photo: Peter Hollamby

resource 19), advice for using apparatus, referring to previous experiments (which might include a video clip or photograph) and background theory to the whole group. The technology allows these to be displayed quickly and clearly in order to maximise the time available to carry out the practical work.

Using a cheap digital video camera to display teacher demonstrations live on a large screen is an inexpensive means of allowing a large class to see clearly what is happening. If concerns such as health and safety, apparatus constraints or the cost of chemicals prevent a large scale demonstration, then this technology can allow the use of smaller quantities to be more visible. The same is true for the screen output from data-logging apparatus or a digital microscope.

To make best use of the time available, details of a practical to be carried out in the next lesson can be displayed so that students will be better prepared. This might include a reminder of how to use the apparatus involved using annotated photographs and diagrams on the screen.

RS•C

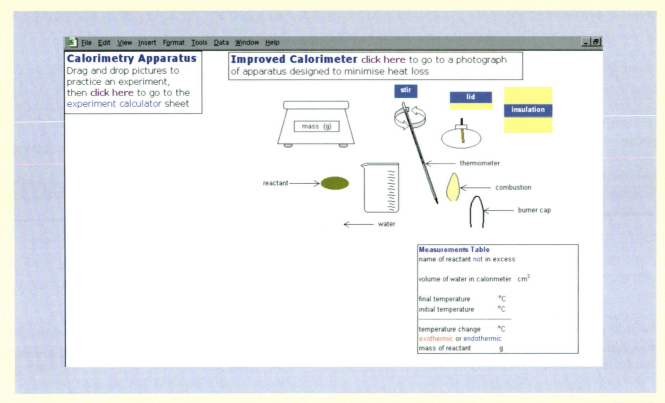

Figure 17
An example (resource 21) where Excel® has been used to create a 'drag and drop' resource for a calorimetry experiment. This can be readily adapted to match a teacher's preferred method. Note the links that can be clicked to open related activities.

Video clips or photographs can be powerful reminders of what was observed whenever experiments are re-visited. Examples of both are included on the CDROM.

Spreadsheet calculators to analyse experimental results

For quantitative experiments, having a computer in the laboratory with a spreadsheet designed to carry out relevant calculations can be a powerful tool. Students enter their measurements and the spreadsheet might calculate:

- each student's result so they know if they've made an error;

- precision and percentage errors; and

- the class average result, error and range.

The whole-class averages can be provided before the end of the lesson so that students can use these as part of their analysis. A print-out of the final spreadsheet may be a useful time saver for the teacher when checking the work students hand in.

An example is included (resource 20) of an Excel® spreadsheet for students to use with a standardisation of hydrochloric acid experiment.

Assembling apparatus

Pictures of apparatus in a Word® or Excel® 'drag and drop' activity can be used in the classroom as a highly visible way for a teacher to demonstrate to the whole class how apparatus should be assembled and to prompt discussion about this.

This resource could be used after or in place of the teacher demonstrating how (or how not) to assemble apparatus with real equipment.

An advantage of using Word® (or Excel®) in this way is that it is straightforward for teachers to amend the page and adapt the pictures and labelling to suit their preferred way of assembling apparatus or the equipment they actually have to hand.

Video clips and still photographs

Short clips of previous experiments are useful to remind students of what they have already seen, particularly in order to focus on the underlying chemistry. Video of industrial examples or reactions that cannot be performed in the school laboratory can be illustrative and stimulating. Examples of both are included on the CDROM.

Using molecular models

Specialist chemical drawing software (for example ACD/ChemSketch, a version of which is available on the CDROM accompanying these guidelines) can be used to create images of a range of chemical structures.

If students are making and manipulating molecular models in the classroom, interactive representations of 3D 'ball and stick' models can be displayed and used to emphasise important features of the molecular models that the students are handling. A wide range of these interactive representations of molecules created using a variety of formats can be found in various internet libraries, readily found by typing phrases such as 'chime' or 'molecular representations' into an internet search engine. A method for creating examples of these from scratch is outlined within the Skills section.

Simply using a photograph of a molecular model and displaying this on a large screen, possibly with annotations, can be useful.

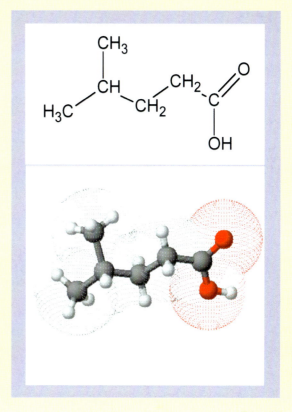

Figure 18
Images representing 4-methylpentanoic acid created using ACD/ChemSketch.

Figure 19

Some views of a MDL Chime molecular representation of 4-methylpentanoic acid that must be viewed within a webpage. These variations may be obtained on the screen using a combination of mouse and keyboard actions.

Wireframe

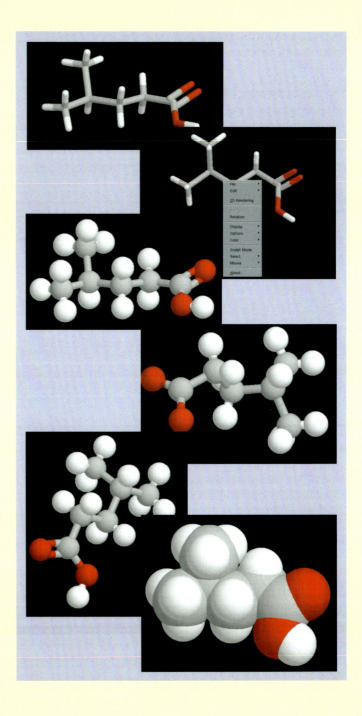

Ball and stick

'Sculpt Mode' (right) allows different parts of the molecule to be moved relative to each other.

Rotation (left) by dragging the image with the mouse.

Spacefill (Van der Waals radii).

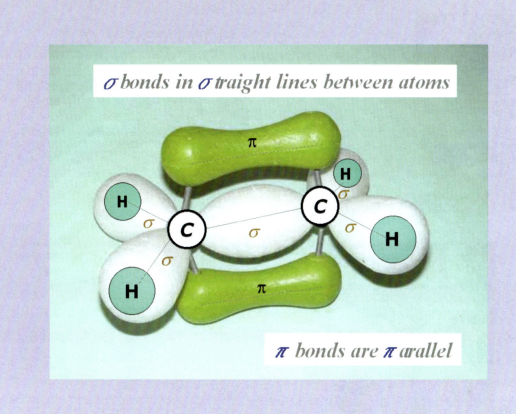

Figure 20
A slide from a Powerpoint® presentation (resource 22) which uses a sequence of annotations of a photograph of PEEL model of ethene to describe the double bond.
Photo: Mike Thompson **http://www.chem-pics.co.uk**

Any of these images can be used to allow students experience of various ways in which chemical scientists use models to aid understanding. They can also be used to illustrate the limitation of models: real atoms and molecules do not look like these images. A quick tour through the stick, ball and stick and space-filling representations can make this clear.

Paper worksheets

If a class is working through a paper based exercise it can be useful to have at least some answers or worked examples available to display on the screen as a means of quickly providing clear feedback. This can be useful to support any learners who are making faster progress than the majority of the group. Such

students can use information on the screen to review their work and then move on without the need for teacher intervention.

This can encourage them to work independently whilst the teacher is free to give their attention to other students. Similarly, the screen might be used to display extension materials for the faster learners.

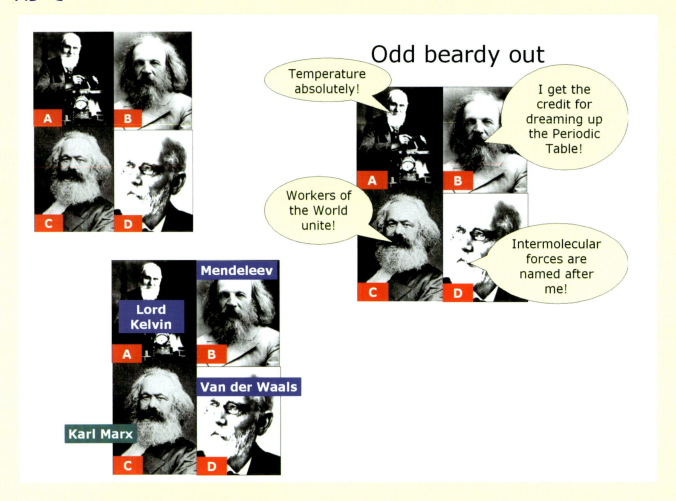

Quizzes

Another 'stimulator' to enliven a lesson. As above, having a bank of these readily accessible on the classroom computer means that they can be introduced into a lesson spontaneously if the teacher decides it might be appropriate.

Figure 21
Three views from a Powerpoint® slide animated 'Quiz' sequence (resource 3).

Review

It may be useful for learners to access any of the ICT materials used in class independently. Being able to use ICT resources that have already been used to teach ideas, skills or facts in class in the learners' own time is repeatedly commented on favourably.

There are several ways of providing learners with access to materials outside of lessons. For example these include making materials available through:

• the school or college computer intranet;

• the school or college website;

• disk or CDROM, the disadvantages of which include that it is time consuming to create numbers of these and updates require more disks to be copied;

• the use of electronic mail; and

• free internet 'user groups' such as those hosted by Yahoo and MSN (Microsoft).

For most of these to be effective the teacher needs to be able to rely on technical support systems in the school or college being effective. The last option above, the internet 'user group', is a useful alternative if technical support is limited or unreliable.

Figure 22

An Excel® spreadsheet (resource 23) that uses various 'forms' to provide feedback. Thanks to Mike Thompson of Uppingham School.

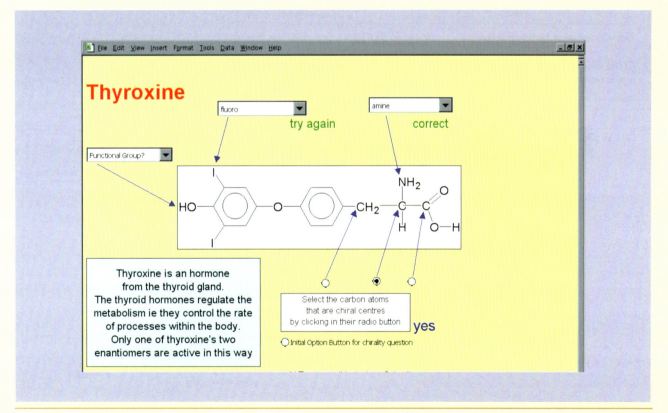

RS•C

Plenary

At the end of each lesson the essential ideas and learning outcomes from the session might be displayed and reviewed with the class. They are saved in a format which means they can be brought up at a later date or accessed independently.

Quick tests / exam question practice and answers

The ease with which materials can be opened and displayed on the screen means that the technology makes it very efficient to display a quick test and then put up the answers for students to mark their own work.

Figure 23

Two linked Powerpoint® slides of an exam question and model answer (resource 24). These are interconnected through a mouse click, illustrating a simple use of the software.

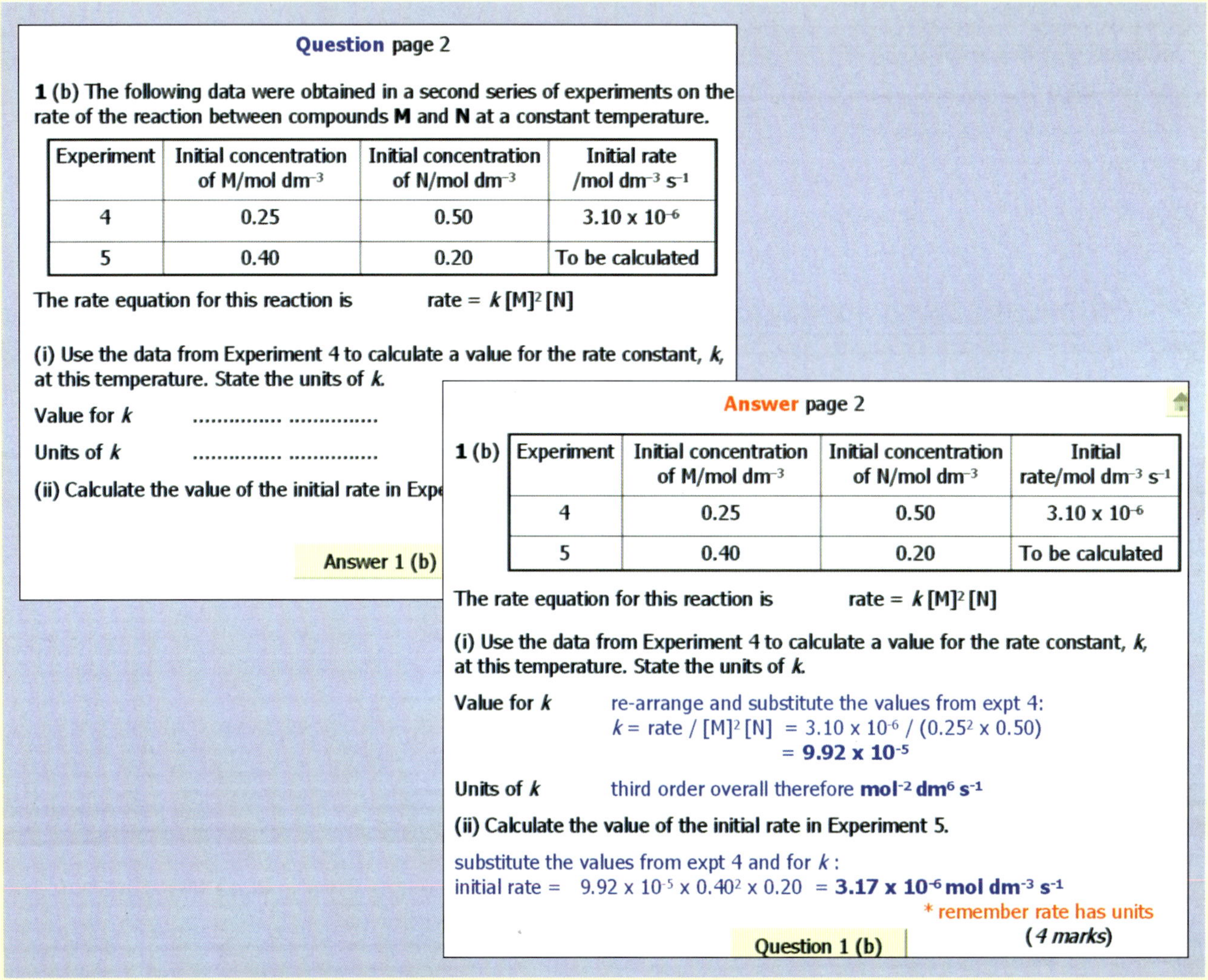

Question page 2

1 (b) The following data were obtained in a second series of experiments on the rate of the reaction between compounds **M** and **N** at a constant temperature.

Experiment	Initial concentration of M/mol dm^{-3}	Initial concentration of N/mol dm^{-3}	Initial rate /mol dm^{-3} s^{-1}
4	0.25	0.50	3.10 x 10^{-6}
5	0.40	0.20	To be calculated

The rate equation for this reaction is   rate = k [M]^2 [N]

(i) Use the data from Experiment 4 to calculate a value for the rate constant, k, at this temperature. State the units of k.

Value for k ....

Units of k ....

(ii) Calculate the value of the initial rate in Exp

Answer 1 (b)

Answer page 2

1 (b)
Experiment	Initial concentration of M/mol dm^{-3}	Initial concentration of N/mol dm^{-3}	Initial rate/mol dm^{-3} s^{-1}
4	0.25	0.50	3.10 x 10^{-6}
5	0.40	0.20	To be calculated

The rate equation for this reaction is   rate = k [M]^2 [N]

(i) Use the data from Experiment 4 to calculate a value for the rate constant, k, at this temperature. State the units of k.

Value for k re-arrange and substitute the values from expt 4:
k = rate / [M]^2 [N] = 3.10 x 10^{-6} / (0.25^2 x 0.50)
          = **9.92 x 10^{-5}**

Units of k third order overall therefore **mol^{-2} dm^6 s^{-1}**

(ii) Calculate the value of the initial rate in Experiment 5.

substitute the values from expt 4 and for k :
initial rate =   9.92 x 10^{-5} x 0.40^2 x 0.20 = **3.17 x 10^{-6} mol dm^{-3} s^{-1}**

                    * remember rate has units
                    (*4 marks*)

Question 1 (b)

Homework and deadline record

The homework record for the group, which may be part of the notice board screen to start the lesson off, can have new entries agreed with the class and typed in at the same time, for instance:

'A presentation introducing new ideas should be reviewed on the school intranet where a web link shown in class can be followed up. An electronic worksheet should be completed and printed out to hand in next lesson.'

If this record is saved where it is accessible to learners outside of lessons it can be a useful reminder or for absentees to find out about missed work.

Electronic worksheets

Several different types of 'e-worksheet' were developed for the project to support learners in reviewing their work. Technology provides the opportunity to extend the role of a traditional paper based worksheet enormously. For example:

- they can be made stimulating and engaging eg through the use of colour, photographs, video clips, animations etc;

- highly structured activities can be created to maximise the support given to learners, perhaps through prompts or clues appearing on demand;

- feedback can be provided automatically in response to student choices eg correct answers, individual comments for each response in a multi-choice question, scores;

- activities can be designed to allow students to pick their own route through eg by providing optional links to follow;

- activities can allow items to be re-organised on the screen eg by 'dragging and dropping'.

The proviso that 'ICT should only be used if it enhances the learning experience and not for its own sake' is especially pertinent here. If a paper or textbook exercise achieves the same end, and as effectively, then continue using these proven methods. As long as the predominant form of external assessment remains written examination students will need to practise completing exercises requiring written answers.

An advantage of the electronic worksheets given here is that they enable teachers to provide a differentiated range of learning opportunities tailored to different abilities, needs and learning styles. Many of these materials are similar in their aims to commercial products. However, providing them as Microsoft Office® application files allows teachers to edit content and create new versions. It also means that the activities should be more readily transferable to be used outside of the school or college.

However, the technical skills required to create some of these, especially the Excel® spreadsheet examples, go beyond the 'comparable to what is needed to produce basic word processed documents' benchmark. In order to address this, the spreadsheet-based e-worksheets described below are available as template files, where all a teacher needs to do, to create their own version, is enter text into highlighted spaces.

Several examples of the project materials reveal answers as the learner works through an activity. This means that it is possible for students to progress through them with little thought, getting the answers as they go and so limiting what is to be gained from the experience. Feedback to the project suggests that in fact few learners will follow such a passive approach, particularly if their classroom experience encourages 'what happens next?' thinking before revealing answers.

Some of the e-worksheets suggested below do not provide learners with automatic feedback following a response. Students can print their draft answers out or save an electronic copy. Colleagues are using a variety of ways to provide feedback to students. These include:

- providing an answer sheet for the students to mark their own work, either on paper or as a computer file;

- displaying answers on a big screen in the classroom and going through these as a class exercise;

- teacher marking of either the printout or of the student answer file posted to the teacher electronically.

Teachers can create new worksheets and modify the existing ones in response to classroom experiences. Over several years and, crucially, with monitoring of how students respond, collections of a variety of these worksheets can be developed to provide a resource bank that caters for the range of learning needs within a particular institution.

RS•C

Drop-down menus

These are akin to the standard 'fill in the blanks' paper worksheet but provide a high level of learner support by placing choices (missing words or phrases) to fill the blank at the point they need to be inserted. Answers can be selected by clicking one from the list.

Using this format allows worksheets to be more highly structured than is possible with a handout. Feedback from several sources suggests that using these worksheets as an exercise prior to a formal assessment has led to gains in student attainment.

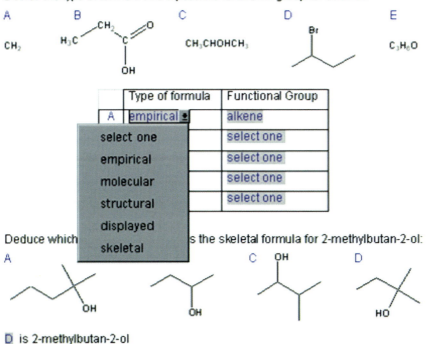

Nomenclature, formulae and isomerism

Decide the type of formula and a possible functional group for each of:

A CH_2

B H_3C—CH_2—C=O | OH

C $CH_3CHOHCH_3$

D (Br)

E C_3H_6O

	Type of formula	Functional Group
A	empirical	alkene
	select one	select one
	empirical	select one
	molecular	select one
	structural	select one
	displayed	
	skeletal	

Deduce which [...] s the skeletal formula for 2-methylbutan-2-ol:

A [structure with OH]

B [structure with OH]

C OH [structure]

D [structure with HO]

D is 2-methylbutan-2-ol

Type the names of the other three structures: A is 2-methylpentan-2-ol

Figure 24

A Word® document (resource 25) that includes drop down menus to make selections from and also text entry forms.

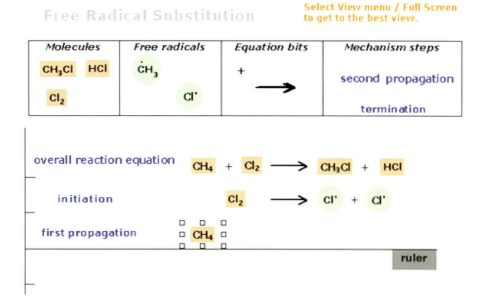

Figure 25
A Word® document (resource 26) that includes pictures and drawing objects that can be re-organised ('dragged and dropped') using the mouse.

Drag-and-drop

These worksheets (which can be produced within Excel® spreadsheets as well as in Word® documents) provide all of the required information and learners are asked to re-organise this on the page. The moveable data can be in any format used in word processed documents *eg* blocks of text, drawings, pictures etc. These can be moved around the page using a mouse or other interactive device. This approach is often used in commercial packages. It is rather clumsy in Word®, but has the advantage that specialist software is not required and the files are more readily transferred as a result.

In terms of differentiation, these worksheets can be designed to provide a lower level of support than the drop-down examples above. For instance all the necessary information to solve a problem can be made available, but in a disorganised way. This can be useful where learners are in the habit of introducing inappropriate items *eg* the occurrence of H· free radicals in alkane reactions with halogens in the example illustrated above.

As has already been discussed, 'drag and drop' activities can also be effective for demonstration or student participation activities within the classroom, with the file projected on a large screen.

In contrast to the high level of support which can be provided through the drop-down menus activities referred to above, this type of e-worksheet provides no support and requires good understanding of a topic. Students apply their word processing skills to first highlight errors and then produce a corrected version of a document containing various textual and diagrammatic errors.

Figure 26
A Word® document (resource 27) containing various errors. Students can apply their word processing skills to produce new versions of this document, first with errors highlighted (below) and then with the errors corrected.

Highlight and correct the errors

Bonding, structure and properties in NaCl, H_2O and the diamond form of carbon

Sodium chloride has a high melting point because it has strong ==intermolecular== bonds. These bonds are strong because the chloride ion has a high ==electronegativity== and this leads to a strong attraction for positive sodium.

For an electrical current to flow through a substance it must have charged particles that are free to move. When sodium chloride is melted it will conduct electricity because its ==electrons== are then free to move.

Water is an unusual substance because of hydrogen bonds between separate H_2O molecules. These are between the ==2-== charge on an O atom of one molecule and the ==1+== charge on one H atom of another.

This partial charge difference arises because O has ==lone pairs of electrons== and ==these== pull the electrons in each O-H bond towards the O atom. This results in the formation of an ==induced== dipole across each bond.

Hydrogen bonding in water:

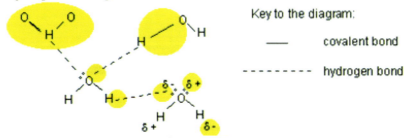

Key to the diagram:

———— covalent bond

- - - - - - - - hydrogen bond

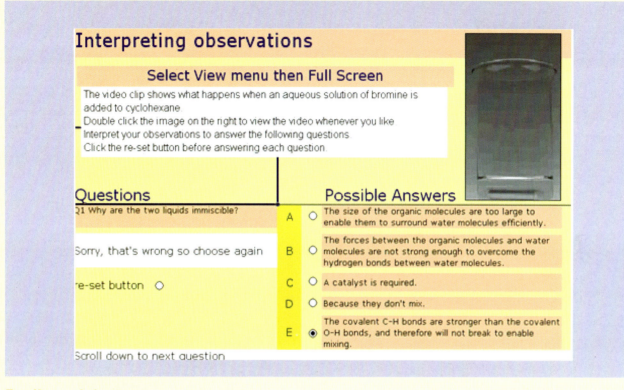

Interpreting observations

Select View menu then Full Screen

The video clip shows what happens when an aqueous solution of bromine is added to cyclohexane.
Double click the image on the right to view the video whenever you like
Interpret your observations to answer the following questions.
Click the re-set button before answering each question.

Questions

Q1 Why are the two liquids immiscible?

Sorry, that's wrong so choose again

re-set button ○

Scroll down to next question

Possible Answers

A ○ The size of the organic molecules are too large to enable them to surround water molecules efficiently.

B ○ The forces between the organic molecules and water molecules are not strong enough to overcome the hydrogen bonds between water molecules.

C ○ A catalyst is required.

D ○ Because they don't mix.

E ⦿ The covalent C-H bonds are stronger than the covalent O-H bonds, and therefore will not break to enable mixing.

Excel® e-worksheets

These can allow learners to make choices and then the worksheet itself provides feedback. As creating these requires a higher level of skill than many of the other project resources templates are provided. These only require text entry and the ability to clear data and formats from cells. Guidelines for creating these from scratch are also provided within the Skills section.

It is possible to hide the familiar features of a spreadsheet (such as the row and column headings, gridlines and sheet tabs) so that the screen is cleared for the learning materials. It is also possible to password protect the areas of the spreadsheet that control the feedback to students. Comments from colleagues through the project year suggest it is preferable not to do this as it makes the process of editing or creating resources over-complicated for many teachers. It was also felt that a student curious about the way the worksheet had been constructed might benefit from being able to explore its workings.

Figure 27

Images, including video clips, may be inserted within Excel® spreadsheets to support activities on the same sheet. This example has a video clip of a reaction accompanied by multiple choice questions and was created from a template (resource 28). Thanks to Ian Bridgewood. Video clip: Mike Docker.

Option buttons

These allow choices to be made, each of which can prompt automatic feedback.

For instance, multiple choice questions can set up to provide an individual comment for each response to each answer, so that comments can be provided that precisely target known obstacles to learning.

RS•C

These use drop-down menus similar to those used in the Word® example above, but within Excel® these can act as prompts leading to automatic feedback.

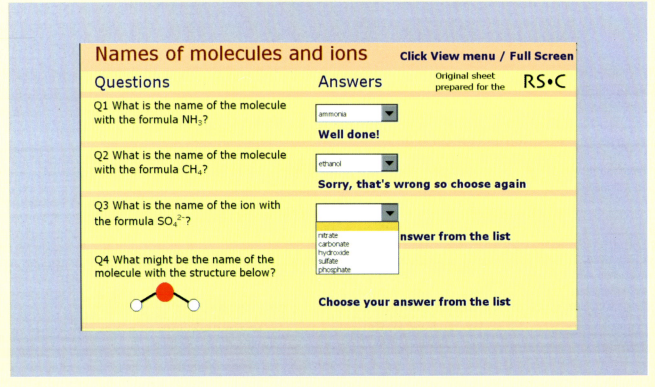

Figure 28
An Excel® spreadsheet (resource 11) illustrating how drop down menus might be used to provided feedback. This was created from a template (resource 30).

Accurate use of terminology

These provide responses to text typed by students. They can be used to encourage to students to use technical terms correctly *eg* misspelling: flourine for fluorine, Fl for F, keytone for ketone, or the wrong context: *eg* atom for ion.

Combining the spreadsheet tools

The previous examples all have a formal 'list of questions' structure. It is just as straight forward to create looser problem solving structures using the same tools. For instance, questions might be arranged around an appropriate diagram or photograph inserted within the spreadsheet. Figure 22 above, a spreadsheet quiz about thyroxine (resource 23), is an example of this.

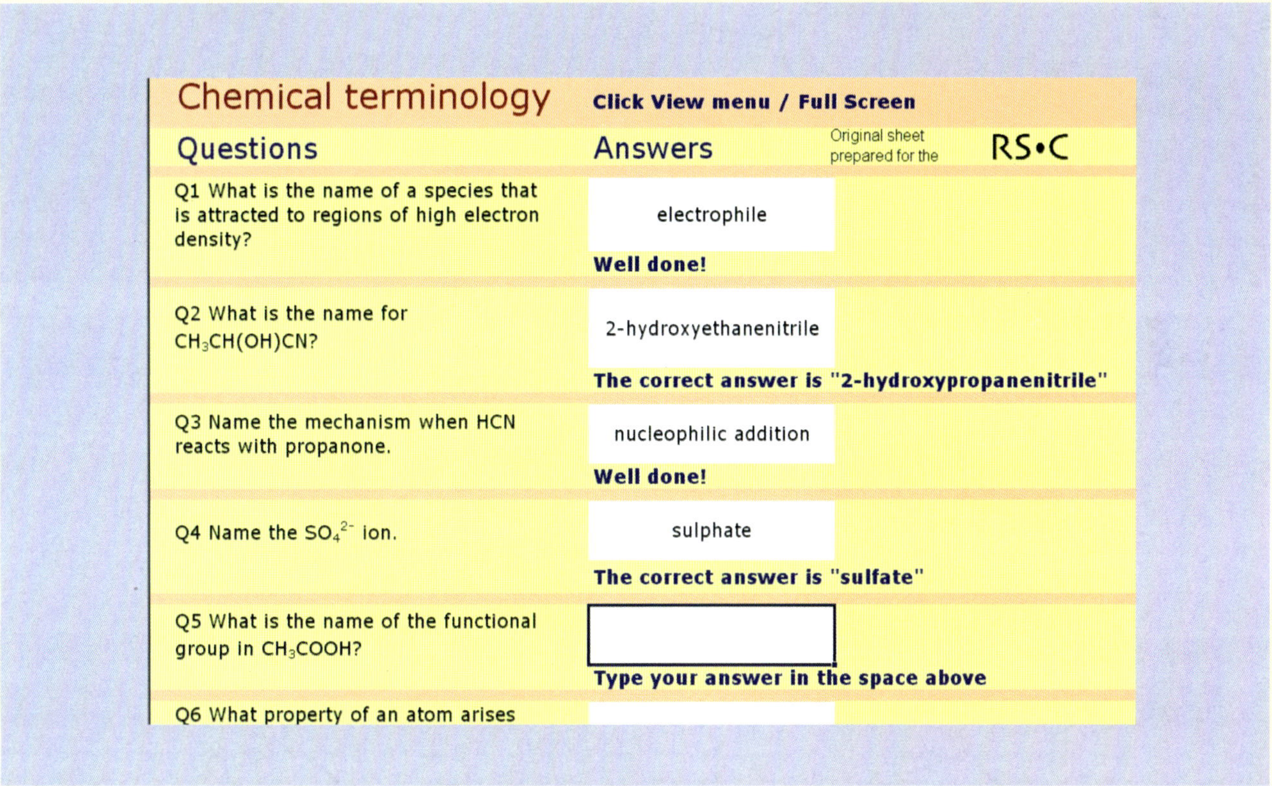

Chemical terminology	Click View menu / Full Screen	
Questions	**Answers**	Original sheet prepared for the **RS•C**
Q1 What is the name of a species that is attracted to regions of high electron density?	electrophile **Well done!**	
Q2 What is the name for $CH_3CH(OH)CN$?	2-hydroxyethanenitrile **The correct answer is "2-hydroxypropanenitrile"**	
Q3 Name the mechanism when HCN reacts with propanone.	nucleophilic addition **Well done!**	
Q4 Name the SO_4^{2-} ion.	sulphate **The correct answer is "sulfate"**	
Q5 What is the name of the functional group in CH_3COOH?	**Type your answer in the space above**	
Q6 What property of an atom arises		

Figure 29

An Excel® spreadsheet (resource 31) illustrating how students might be supported in writing technical terms accurately. This was created from a template (resource 32).

Powerpoint® e-worksheets

These can be designed to pose questions or problems to which the answers are revealed on the next mouse click or key press. The process can be made more subtle by revealing a sequence of prompts before the answer so that the learner's thought process can be guided. They can incorporate photographs, diagrams, movie clips as illustrated in the examples of presentations above.

Links to other Powerpoint® slides, web pages or Word® documents can be added. These allow a teacher to build in choices for the routes taken through an activity. This might be used to select level of difficulty or to allow revision of terms used in a question, but only if the learner decides that they need to do so.

Figure 30

Part of an animation sequence from a Powerpoint® e-worksheet (resource 33) that guides learners through questions about a molecular structure. This includes a range of techniques including direct questions, prompts and screen highlights that can appear when needed and then disappear. Templates for creating this type of e-worksheet are provided (resource 34).

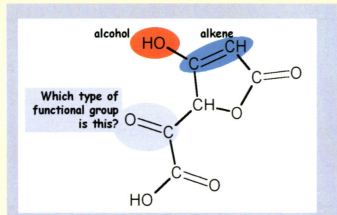

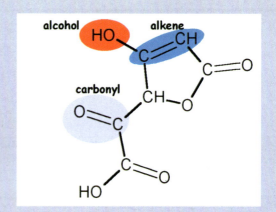

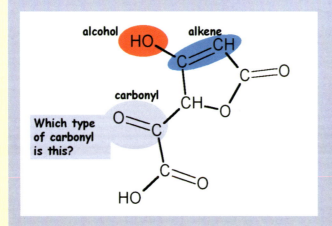

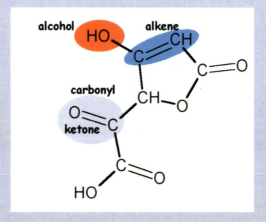

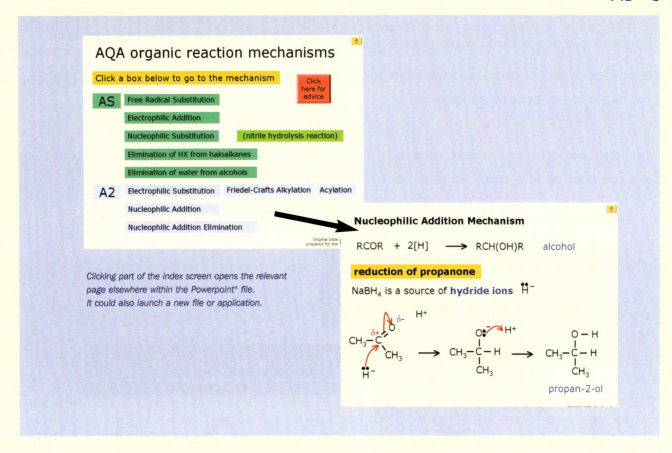

Clicking part of the index screen opens the relevant page elsewhere within the Powerpoint® file. It could also launch a new file or application.

Powerpoint® tutorials

Links within a Powerpoint® presentation enable learners to plot their own route through the content. A popular request for development during the project was a collection of animated organic reaction mechanisms. These are organised with an index page with links that allow learners to select a particular reaction but then return directly to the index page from there.

Between them, resources 6, 7 and 8 contain a complete collection of the organic reaction mechanisms required for most 16–19 courses. For the project they have been published in a format best suited for end of course revision. However, because they have been created using Powerpoint®, teachers of chemistry have the option of adapting this material in several ways.

Figure 31

An index slide for an organic reaction mechanism tutorial (resource 6). This contains links to other slides within the same Powerpoint® file. Note the 'home' button in the top right corner, which links back to the index screen.

For instance:

- copying the slides for a particular topic and using these for the teaching and review of these,

- editing the content and animation order of the mechanisms to suit personal preferences or other examination specifications, and

- using material here as templates to create examples with different formulae to include in review questions.

Powerpoint® can be used as a medium for linking into a coherent whole several different activities, including those created within other applications. Consequently it can be used to develop tutorial packages that might combine several of the features discussed in the previous sections and provide a structure that can be more easily accessed independently by students.

Figure 32

An example of a Powerpoint® file (resource 5) describing an application of chemistry that students are unlikely to meet elsewhere. It relates this context to pre-16 curriculum content and also follows an e-worksheet 'what happens next?' approach and can be used with the whole class or as an independent tutorial.
NPL Super Black images, used with permission of NPL
Road signs, used with permission of HMSO
Space images courtesy of NASA / STScI

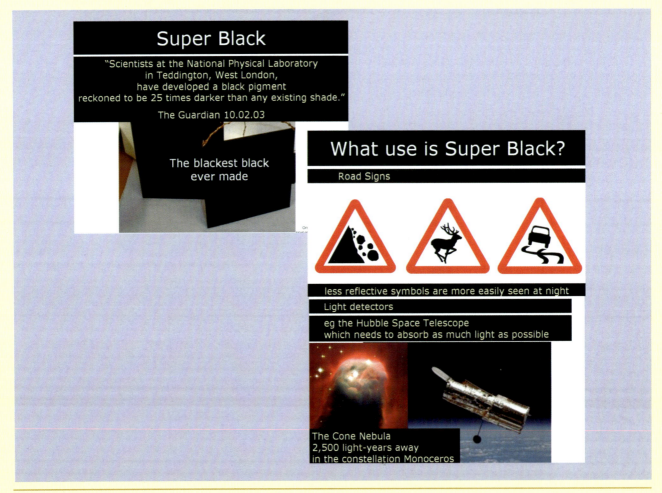

Other resources

The previous examples are distinguished by the fact that they can be used, created, edited and distributed using software that is very widely available.

There is a wide range of pre-prepared resources available to support chemistry teaching to all ages. Much of this is designed to be accessed by students independently or working in small groups rather than used within whole-class teaching. Furthermore it is rarely possible to customise the materials by removing irrelevant, incorrect or misleading information.

However, it is significant that this material is often provided in a form that, although customisation is not possible easily, allows small units to be 'dropped into' a scheme of work in the same way as implied for the largely Microsoft Office® based materials described previously. For instance, it might be possible to use a pre-prepared web page, or to copy an animation or video clip out of a much larger resource and use these to support a lesson activity or e-worksheet for independent study.

Examples, including Royal Society of Chemistry CDROMs and websites for teachers, Java Applets, Macromedia animations and web pages using JavaScript are described in the next section, Types of ICT resource.

Technical support

To be able to use ICT, either in the classroom or to support independent study activities, teachers need to rely on the computer hardware to work and the relevant files or applications to be available on demand.

In most cases it is unreasonable or unrealistic to expect busy teachers to be responsible for providing and maintaining the necessary technical services.

Establishing good relations with the people responsible for delivering technical support services can be a crucial factor in making ICT materials available for learners reliably.

What happens if there is a technical hitch?

There are any number of potential technical failures eg the computer crashes, a data projector bulb blows, the internet connection is lost or becomes too slow, there is a power failure.

Fear of these possibilities happening in front of learners is often quoted as a reason why colleagues are reluctant to develop their classroom ICT skills.

Any lesson plan that makes assumptions about what is possible in the time available, for example the presence of a laboratory technician, apparatus and chemicals being available as requested, student behaviour etc needs to have an alternative in mind in case of the unforeseen. It is sensible therefore, as with any other lesson, to have a 'Plan B' in mind for any lesson involving ICT: something not reliant on technology or other material resources. For instance a quick paper classroom activity based on a simple question allows time to collect thoughts followed by a board activity. Accepting that these things happen and dealing with them is all part of continuing to develop resourcefulness.

It is certainly preferable to abandon the original plan rather than spend an unpredictable amount of time floundering around in front of an increasingly restless class trying to sort it out. The dynamic ICT lesson originally planned will still be there for a later date.

RS•C

Types of ICT resource

Many of the types of resource referred to below were introduced in the previous section to illustrate 'Integrating ICT into a sequence of chemistry lessons'.

This 'Types of ICT resource' section provides a different perspective by cataloguing resources according to the type of ICT application used to create them. It is designed to be read independently, so there is some inevitable overlap with the previous section, although different examples have been chosen.

Comments from a chemistry teacher who has been experimenting for some time with making use of ICT with his students:

'I would like to see short, snappy resources that I can integrate easily into my teaching, something that does not control the direction of my lesson.

I want resources that enable students to visualise chemistry in novel ways.

I don't want a set of notes on a web-page unless it is effectively integrated with animation/video and instant feedback questions.'

The emphasis is on activities that make use of the facilities available to the majority of teachers of chemistry and requiring a skill level no higher than experience suggests the majority have attained.

Of course, in reality there is a wide diversity between teachers of chemistry in access to facilities and skill levels, however experience suggests that the focus chosen is appropriate for the majority and it is hoped that these materials offer something of interest to the rest.

A sensible approach, particularly if relatively inexperienced in this area, would be to take one idea for a resource, use it and assess its effectiveness rather than risk being overwhelmed through attempting too much at once. The 'Skills' sections of these guidelines provide outline instructions for creating many of the resources described below.

Why focus on Microsoft Office® applications?

Microsoft Office® applications are available and familiar to the majority of teachers and their students. Although the software is regularly updated, the differences are relatively minor, so any materials produced using Office applications are likely to be widely accessible and transferable for some time.

Quick glance: some uses for Microsoft Office® applications

PowerPoint®
- Display information on a large screen in the classroom: sequences of text, photographs and diagrams, video clips and animations
- Provide a framework for seamless links to other applications: eg internet, video, CDROM, Word®, Excel®, molecular simulations
- Display and link information on single computers: eg electronic worksheets which pose questions and can provide hints before revealing an answer.

Word®
- Create 'drop-down-menu' worksheets which allow choices to be made about what text is entered at specific points in the document.
- Create 'drag and drop' activities which allow objects on the page to be re-organised. Useful with whole class or as a worksheet.
- Edit prepared text eg identify and correct the errors in a document.
- Create basic web pages: for teachers to create structured paths to link resources, making them more accessible, for student projects and to publicise student achievement.

Excel®
- Represent numeric data graphically.
- Create models for the investigation of the effect of variables
- Create electronic worksheets that can provide responses and scores.
- Carry out calculations, including those needed to support practical work such as assessment of errors.
- As with Word®, create 'drag and drop' activities which allow objects on the page to be re-organised. Excel® is useful as other types of activity can be included on the same worksheet.

Because these resources are created using Office® they are all accessible for editing, copying and as templates for completely new activities.

It is commonplace now for chemistry teachers to produce basic word processed documents for their students and the level of skill required to do this is the benchmark used within the project when suggesting possible activities.

Consequently, the main focus here is to suggest some ways in which chemistry teachers might make more effective use of the facilities available to them and to do so through developing their current level of skill.

PowerPoint® presentations: ppt files

PowerPoint® is a versatile means of displaying information clearly and of interacting with it. It can also be used to link a range of different ICT applications and materials to form an 'electronic' lesson. For instance, a sequence of PowerPoint® slides might include links to a Word® document, a live website and a specialist piece of software so that the teacher in class or a learner working independently is able to move fluently between all of these and weave them into a varied but coherent learning experience.

Editing and sampling

Many teachers comment they have found that rather than preparing 'slides' for PowerPoint® ppt files as fixed resources to be used in the same form for several years, they in fact continually amend and add content as they use them. Familiarity follows from practice, so that the process of modifying and updating these slides can become quick and straightforward.

It is rare for any two teachers to deliver a topic in exactly the same way. An advantage of being able to adapt an existing PowerPoint® file is that colleagues' work can be customised to suit another teacher's needs or style.

For example, the project materials include animations of organic reaction mechanisms (resources 6, 7 and 8).

Figure 33 (opposite)
The following two slides relating to electrophilic addition are from a PowerPoint® file (resource 8) summarising all the organic reaction equations and mechanisms covered during a post 16 course. The information is revealed point by point using the mouse or keyboard.

Each file is a collection of post-16 reactions for all the organic topics. Teachers might find it useful to copy the relevant slides from one of these to create a new file to support an individual topic. Furthermore, some teachers might prefer the mechanism steps to be annotated with brief explanations or descriptions. Slides displaying photographs or video clips of the actual reactions might be added. In this way a new resource can be developed that more closely matches the way an individual teacher operates or the expectations they may have of their own students.

Any of the text or diagrams may be amended so that the original resource can be adapted to suit a particular approach to learning. Figure 33 illustrates this, and the lower row of slides were readily produced by editing the two original slides shown above. The background colour and some text have been amended. Explanatory notes that appear and then disappear as the presentation is advanced have been included. This adaptability is a strength of using generic software, such as PowerPoint®, to create teaching materials.

Another possibility, allowed by the ease with which existing presentations can be adapted, is using these as templates. Often, applying the find and replace option in the Edit menu allows new versions of an existing file to be created quickly. For example a quiz template may be used to generate new questions to support other topics.

Images and animations

Being able to display a dramatic or intriguing image on a large screen can be a powerful tool in the classroom. PowerPoint® allows this and also has the capacity to add text and other annotations to the image.

Figure 34 (above)

A PowerPoint® version of a 'Family Fortunes' style game (resource 35). This may be used as a template to produce new versions of the same activity eg through using find and replace. Thanks to Peter Hollamby.

Figure 35 (right)

A PowerPoint® sequence (resource 36) which uses annotated diagrams and photographs of experiments to link practice and theory. Photo: Peter Hollamby.

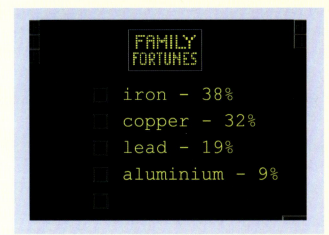

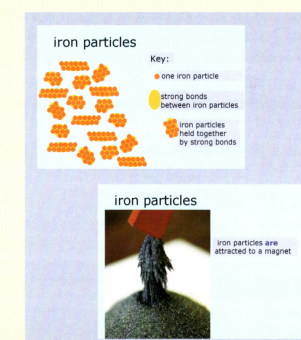

RS•C

The project resources include a small collection of copyright free images, including photographs, video clips and animated diagrams. There are also links to internet sources of others. These can be included within non-profit educational resources if the source is credited.

Video clips can be inserted within a PowerPoint® presentation, so that the clip can be played as part of the sequence of events displayed. The clips may be 'embedded' or run within media playing software that is linked to the PowerPoint® slide. The clips can be displayed so that they may be played repeatedly or individual frames can be frozen and examined more closely.

Word® documents: doc files

ICT provides the opportunity to extend the role of a traditional paper based worksheet enormously through the use 'electronic' worksheets, referred to within the project resources as 'e-worksheets'.

Drop-down menus

These make use of a type of 'Form' that creates a list of choices that appears on the screen at a specified point. When one of these is selected the word or phrase appears on the page at that point and the form itself disappears from view. Other Forms that can be placed within a document include text entry, where any text can be typed and check boxes, which allow choices to be identified. When the Forms are activated the rest of the document is protected and cannot be changed.

E-worksheets using Forms are very quick and easy to produce, as described within the Skills guidelines. They allow the creation of activities with high levels of structure and focused support for learners.

Feedback from several sources suggests that using these (and other types of e-worksheet) with students preparing for assessments of knowledge can have a positive effect on attainment.

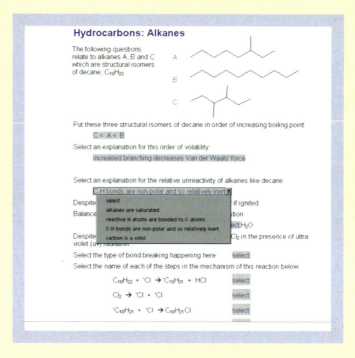

Figure 36
A Word® 'drop down menu' document revision activity (resource 37).

Drag-and-drop

This feature can be used for interactive classroom displays and also independently for e-worksheets. Students can be asked to re-organise information displayed on the screen. The moveable data is in the form of text boxes or drawing objects which can be moved around the page using the mouse.

A drag and drop exercise might provide all the required information but in a disorganised manner. This allows the development of activities that have less structure and support than, for example, that provided in the previous 'drop down menu' exercises.

This approach is often used in commercial packages. It is rather clumsy in Word® (and in Excel®, which can also support this). The Skills guidelines describe the issues involved and suggest ways to create e-worksheets with this feature.

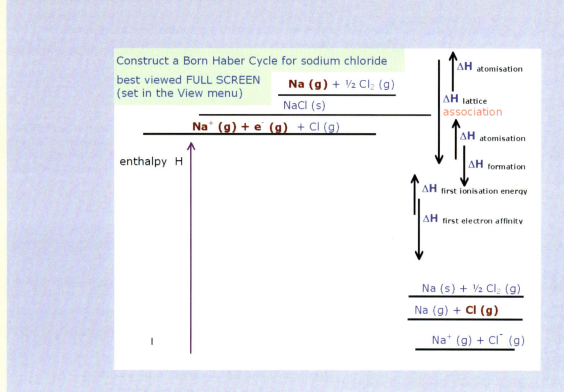

Figure 37

*A Word® 'drag and drop' document (resource 38). The cycle may be
assembled using the mouse to re-arrange the various images. An
advantage of using this format is the opportunity it provides for teachers to
amend the text to suit their own teaching style and also to produce cycles
for other substances.*

Edit existing text and images

Students can be provided with a document containing errors. They can identify the errors by highlighting them in some way and save a copy. They can then produce a second copy with the errors corrected. Both copies can be printed out.

Alternatively, an entirely electronic approach is for the student to correct the errors and then save the updated document to somewhere the teacher has access (eg to disk, a network folder, e-mail etc).

Word® has a facility to automatically compare two documents (Tools menu, track changes, compare documents). This may be utilised to compare the student version of the document with the original and automatically highlight any differences, which can help to make the marking process more efficient.

Figure 38

A Word® document (resource 27) which contains deliberate errors and a student's version where these have been corrected. The compare documents option has been selected to automatically highlight the student's amendments. This type of activity can be structured to provide minimal support, one example from a range of responses, possible using ICT, that can be employed to match diversity in student need.

Correct the errors

Bonding, structure and properties in NaCl, H_2O and the diamond form of carbon

Sodium chloride has a high melting point because it has strong intermolecular bonds. These bonds are strong because the chloride ion has a high electronegativity and this leads to a strong attraction for positive sodium.

For an electrical current to flow through a substance it must have charged particles that are free to move. When sodium chloride is mel it will conduct electricity because its electrons are then free to move.

Water is an unusual substance because of hydrogen bonds between separate H_2O molecules. These are between the 2- charge on an O atom of one molecule and the 1+ charge on one H atom of another.

This partial charge difference arises because O has lone pairs of electrons and these pull the electrons in each O-H bond towards the atom. This results in the formation of an induced dipole across each bond.

Hydrogen bonding in water:

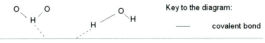

Key to the diagram:

—— covalent bond

Bonding, structure and properties in NaCl, H_2O and the diamond form of carbon

Sodium chloride has a high melting point because it has strong ~~intermolecular~~ionic bonds. These bonds are strong because the chloride ion has a ~~high electronegativity~~single negative charge and this leads to a strong attraction for positive sodium ions.

For an electrical current to flow through a substance it must have charged particles that are free to move. When sodium chloride is melted it will conduct electricity because its ~~electrons~~ions are then free to move.

Water is an unusual substance because of hydrogen bonds between separate H_2O molecules. These are between the ~~2-~~delta negative (δ-) charge on an O atom of one molecule and the ~~1+~~delta positive (δ+) charge on one H atom of another.

This partial charge difference arises because ~~O has lone pairs of electrons and these pull~~an O atom has a higher electronegativity than an H atom and so pulls the electrons in each O-H bond towards the O atom. This results in the formation of an ~~induced~~permanent dipole across each bond.

Hydrogen bonding in water:

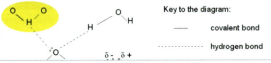

Key to the diagram:

—— covalent bond

- - - - - - - hydrogen bond

Create web pages: html or htm files

Word® documents can very easily be saved as web pages. Being able to do this can be convenient for teachers or students who haven't either the expertise, inclination or time available to make use of specialist 'web page authoring' software such as Microsoft Front Page®.

Uses of web pages to support learning include:

- Providing frameworks of linked web pages that guide students to learning resources.

  This can be useful if students are accessing resources independently eg on disk, a website or a computer network. A framework of pages is preferable to simply providing the files, even if these are saved within appropriate folders, as it allows learning paths to be created and instructions for how to use the resource included.

- Displaying interactive molecular representations.

- Free software (eg Chime or RasMol) can be downloaded from the internet which allows various choices to be made regarding the molecular image. These include being able to rotate the view, changing the type of molecular structure displayed and altering the size of the image. An example is shown in Figure 39.

- To support other types of interactive chemistry learning material that run within web pages eg JavaScript features, Shockwave animations or Java Applets, which are all described below.

- As a useful means of saving student work and providing evidence of their achievement and progress. The web page format provides a highly effective way of showcasing achievement, storing or sharing work. For instance, web pages are ideal to make students work available across a computer network for others to see. Print outs of some examples can be included in poster displays around classrooms and corridors. Both approaches can make it explicit that ICT is an integral component of the students' science course.

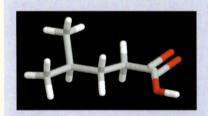
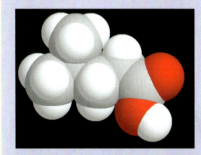
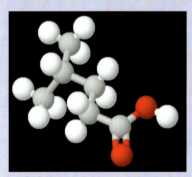
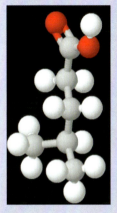
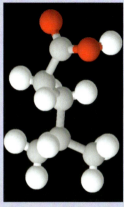

Figure 39

Some views of a MDL Chime molecular representation of 4-methylpentanoic acid that must be viewed within a webpage. These variations may be obtained on the screen using a combination of mouse and keyboard actions.

RS•C

Excel® spreadsheets: xls files

Excel® provides many opportunities for creating resources and activities to support teaching and learning chemistry. These resources do not need to have the traditional 'gridlines' view of a spreadsheet and they can include images, both still and video. Some possibilities, discussed in more detail below:

• graphs and charts of trends and patterns;

• models that allow the effects of variables to be investigated;

• questions and quizzes with automatic responses and scores;

• analysis of experimental results, including errors;

• 'drag and drop' activities.

Graphs and charts

Most students are likely to have learnt the basic skills needed to create graphical representations from data sets using Excel®. However they are likely to need significant support in applying these skills to a chemical problem, for example with choosing an appropriate format for plotting the points and with labels and scale. This can turn a comparatively trivial process into a difficult activity to manage with large class sizes and can become a lesson where the chemistry content becomes secondary to teaching students ICT skills.

This has potential as a relatively quick way for students to investigate trends and patterns of behaviour, but the teacher needs to establish the level of ICT skill and understanding of the group and have a clear strategy at the outset to deal with any shortcomings.

Displaying Excel® to the whole class using a large display screen and going through the steps involved in creating a graph or chart can be a time-effective way of reinforcing the thinking involved. This also supports students learning to hand draw versions of these.

Models

The project resources include some Excel® spreadsheet models of chemical relationships that allow the effect of changing variables on a relationship to be investigated.

Figures 40a (below) and 40b (opposite)
Views of two linked Excel® sheets that allow the enthalpy change for a calorimetry experiment to be calculated and the effects of heat losses modelled using 'slider' controls (resource 21).

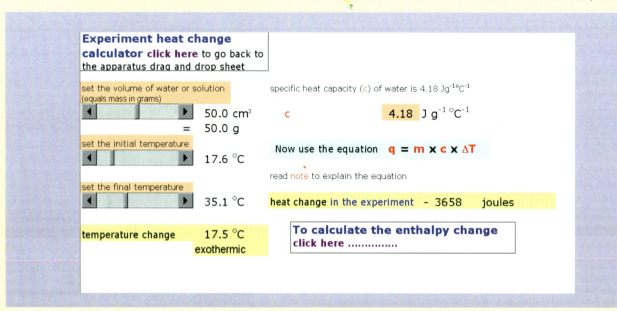

Experiment heat change calculator click here to go back to the apparatus drag and drop sheet

set the volume of water or solution (equals mass in grams)

50.0 cm³
= 50.0 g

specific heat capacity (c) of water is 4.18 Jg⁻¹°C⁻¹

c | 4.18 J g^{-1} °C^{-1}

set the initial temperature

17.6 °C

Now use the equation $q = m \times c \times \Delta T$

read note to explain the equation

set the final temperature

35.1 °C

heat change in the experiment - 3658 joules

temperature change 17.5 °C
exothermic

To calculate the enthalpy change
click here ....

Questions and quizzes

These can allow learners to make choices and then provide feedback. Creating these requires a higher level of skill than many of the other project resources. The project materials include guidelines setting out how these features can be created and, to widen access, templates are also provided. The templates only require text entry and the ability to clear data and formats from cells.

Multi-choice

These use 'Option Button Forms' which allow choices to be made, each of which can prompt automatic feedback. Feedback can be tailored to match each response.

For instance, each choice to a multiple choice question could be chosen to cover common misinterpretations or issues that came up in class with a particular group of students. The automatic response revealed for that choice might discuss the reasons for this being wrong. Used in this way the resource can allow the learning needs of this group to be closely matched and provides targeted feedback that traditional resources cannot. For an example, see Figure 27 in the Integrating ICT section.

Drop-down menus

These use 'Drop-down menu Forms' similar to those used in the Word® example above (Figure 36), but, within Excel®, making a selection can prompt automatic feedback (Figure 28).

Accurate use of terminology

These e-worksheets use logical formulae to provide responses to text typed by students. They can be used to encourage to students to use technical terms correctly eg spelling: flourine, Fl, keytone or the wrong context: eg atom for ion. See Figure 29 for an example.

Calculate the enthalpy change
click here to go back to the calculation of the experimental heat change

heat change in the experiment   - 3658   J   exothermic

Calculating the enthalpy change per mole:

mass of reactant   1.40   g

relative molecular mass of reactant   56.0

moles of reactant   0.025 moles

enthalpy change   =   - 146300 J mol^{-1}
per mole of reactant   =   - 146.3   kJ mol^{-1}

increase insulation to allow for heat losses (assuming negligible measurement errors)

◄ | | ►   25 % of heat was lost to surroundings
decreasing heat losses to the surroundings
enthalpy change   exothermic

adjusted to allow for heat losses   - 194.6 kJ mol^{-1}

Improved Calorimeter click here to go to a photograph of apparatus designed to minimise heat loss

Analysis of experimental results

For quantitative experiments, having a computer in the laboratory, with a spreadsheet designed to carry out relevant calculations, can be a powerful tool. Students enter their measurements and the spreadsheet can calculate:

• each student's result so they know if they've made an error;

• precision and percentage errors;

• the class average result, error and range.

The whole-class averages can be provided before the end of the lesson for students to use as part of their analysis. A print-out of the final spreadsheet may be a useful time saver for the teacher when checking the work students hand in. Resource 20 is an example of this approach.

Drag-and-drop

These can be created in Excel® spreadsheets (Figure 17) in the same way as with Word® document drag-and-drop activities (eg Figure 37). This feature can be used for interactive classroom displays, where the teacher or student move images around on the screen, and also within independent e-worksheet exercises.

Linking different types of activity

An advantage of using Excel® is that several of the features described above may be used within the same xls file to form a combined resource.

Figure 41

Labelled calorimeter photographs within the same Excel® file that contains the drag and drop apparatus diagrams (Figure 17) and enthalpy change calculator spreadsheets (Figure 40), all connected by 'mouse click' links.
Photos: Mike Thompson **http://www.chem-pics.co.uk**

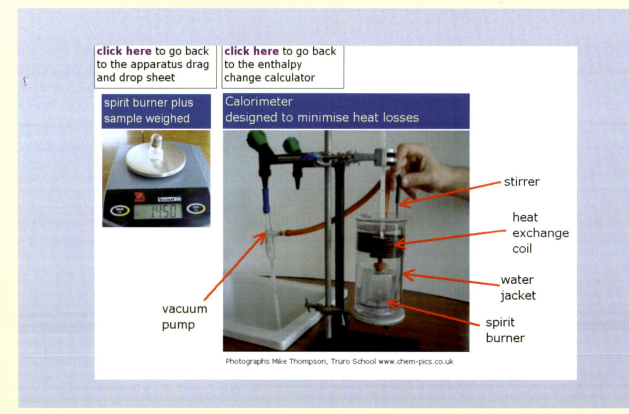

click here to go back to the apparatus drag and drop sheet

click here to go back to the enthalpy change calculator

spirit burner plus sample weighed

Calorimeter designed to minimise heat losses

stirrer

heat exchange coil

water jacket

spirit burner

vacuum pump

Photographs Mike Thompson, Truro School www.chem-pics.co.uk

What other types of ICT resources are available?

There are a range of specialist learning material packages and tools available that support aspects of chemistry teaching.

These guidelines do not discuss commercial products unless these are available free of charge, apart from some resources made available by chemistry teachers who make a charge to cover their costs and development time.

There are an ever increasing number of commercial products available and care is needed when making judgments about their effectiveness to support teaching and learning. In particular care needs to be taken to see through superficial glossy presentations. A benefit of using and creating some of the project materials is that teachers will inevitably become more aware of what makes an ICT resource effective for learners. As a consequence teachers should become more discerning when making judgments about the value of commercial products.

In fact many of the project activities use approaches to learning found in commercial products. An advantage of using Microsoft® applications to create these features is that since the software is generic, students are likely to be able to use the materials on computers around the school or college and at home without the inconvenience and expense of installing expensive specialist software. A disadvantage is that the presentation of the materials won't necessarily be as impressive and there is more likelihood of technical hitches.

The resources available range from tutorials that cover significant areas of content, often in the form of independent study units, to small learning units that can be dropped into any lesson plan or collection of review materials for independent learning. Often they have to be used as supplied. This can mean that to be of value the content and style should closely match teacher and student needs. It is often difficult to amend the content of these packages without some specialist knowledge, which usually includes computer programming skills.

There are several examples of specialist software for drawing chemical structures and diagrams. A version of a free ACD/ChemSketch drawing package is included with the project resources. Chemical drawing software is relatively straightforward to use if some time can be set aside to become familiar with it. This is a worthwhile investment as there are many benefits to chemistry teachers in being able to create simple or complex chemical representations such as molecules or assembled apparatus. Furthermore, when confident with these techniques, it is a relatively simple additional step to be able to convert such molecular images into interactive 3D representations that can be moved and rotated.

Hot Potatoes is free software designed to be relatively easy to use for the creation of learning exercises.

Most ready-made ICT resources for chemistry teaching must be accepted as they have been designed, which is reasonable if they closely match a teacher's needs. In practice the differences between design and fitness for purpose for many of these materials often lead to teachers not using them.

Creating most of the resources outlined below require programming skills and specialist knowledge which are beyond the ability and interest of most chemistry teachers. They are invariably presented to users in a form that forbids editing and customisation.

However, often only a basic understanding is required in order to be able to extract what is useful from their original context and integrate this into a teacher's personal 'electronic resource bank'.

RS•C

Quick glance: ICT resources other than Microsoft Office®

Chemical drawing software
ACD/ChemSketch is included on the CDROM accompanying these guidelines. 2D and interactive 3D representations of molecules can be created. It also has a selection of template drawings, including laboratory apparatus.

Molecular visualisation eg pdb, mol
3D representations of molecules that can be rotated, enlarged and re-configured.

Websites html, htm
There are an enormous number of websites related to chemistry teaching. Discernment is needed to make good use them but they are a rich source of the types of material listed below to include in teaching materials, subject to copyright permission being obtained.

Still images eg gif, jpeg, tiff, bmp
Readily available, particularly via the internet and valuable for enhancing resources, again subject to copyright considerations.

Video clips eg avi, mov, mpeg
Various sources of these for chemists. Of variable quality and large file sizes so if accessed via the internet require a fast connection.

Macromedia Flash (swf) and Shockwave (dir or dcr) files
Produced using authoring software (Flash and Director respectively) and a means of providing interactivity and multi-media within web pages or stand-alone materials. Widely used within educational tutorial packages.

JavaScript
Used when writing web pages to add interactivity eg moveable object, lists of choices. Does not require additional software other than current version of internet 'browser' (the software needed to view web pages). Not to be confused with Java Applets.

Java Applets
Small software applications that can be embedded within a web page. Used to provide interactive features similar to Flash and Shockwave. Not to be confused with JavaScript.

Specialist authoring software
Hot Potatoes: Templates and tools to create web based resources with interactivity and scoring yet requiring no programming knowledge. Free to educationalists if the activities created are made freely available.
SimChemistry: Specialist chemistry software that can be used to create models of particle behaviour, including effects of attractions, kinetic energy and temperature. Free evaluation but user licenses need to be purchased.

RSC publications
CDROMs, paper based and websites. Rich sources of materials and ideas.

Datalogging
Traditional application of ICT for the science classroom. Enables data to be collected quickly so that focus can be on analysis and interpretation. Allows very slow and very fast processes to be investigated. Can enhance teacher demonstrations. Issues with classroom management and expense.

Chemical structure drawing software

There are several software packages available designed to draw chemical structures. Examples include CambridgeSoft ChemDraw and MDL / ISIS Draw and ACD / ChemSketch. Some of these companies make older versions of their products available free of charge.

The Canadian company Advanced Chemistry Development (ACD) has kindly agreed to allow the freeware version of ChemSketch to be included on the project CDROM. Guidelines

Figure 42
a) Various molecular representations of ethyl ethanoate created in the ACD / ChemSketch Structure page:
b) Various molecular representations of ethyl ethanoate copied from the ACD / ChemSketch 3D Viewer, which includes estimates of bond lengths and angles:
c) Examples of structures and diagrams copied from ACD / ChemSketch Templates:

for using this software, which are broadly similar to using any chemical drawing software, are included within the Skills guidelines.

Some of the features that can be created using this version of ChemSketch:

- 2D molecular structures: fully or partially displayed or skeletal;

- 3D representations: wireframe, ball and stick or space filling;

- interactive 3D views of the structures displayed within a viewer included with the ChemSketch software;

- several image templates, including biomolecules and lab apparatus;

- molecular representations can be saved in a form which allows interaction with the structures when these are viewed within a web page.

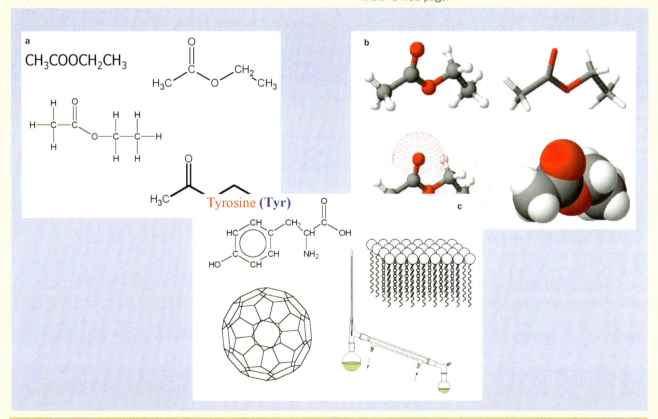

RS•C

Viewing interactive molecular representations

There are an enormous number of copyright free interactive molecular representation files available via the internet, many held within on-line libraries. These have a variety of formats recognisable from the letters at the end of their file names, common examples are pdb and mol.

Specialist molecule file viewer software is needed to display and interact with these files. In most cases this software will run by itself, although it is often linked to a web page.

A detailed discussion of how these files are created or existing ones edited is beyond the scope of this project. However, the Skills guidelines describe how a structure drawn in ChemSketch may be exported as a mol file and then embedded within a web page using the MDL Chime 'plug-in' molecule viewer software.

MDL Chime

MDL Chime is 'plug-in' software that displays 2D and 3D molecules directly within a web page and works with both Netscape and Microsoft® browsers. It will not work without being embedded within a web page. It is a widely used way of displaying 'molecular visualisations'. The molecules displayed on the page are 'live', meaning they are not just pictures of chemical structures, but images that can be rotated, reformatted, and reshaped. The Chime website is at: **http://www.mdlchime.com/chime/**

Chime is free software which can be downloaded from the MDL website. There is a requirement to register and accept a user license agreement. In addition to single use license agreements, MDL also makes selective grants of licenses to customers who want to use Chime over a network or redistribute it to other users. Instructions for how to apply for one of these can be found on the Chime website.

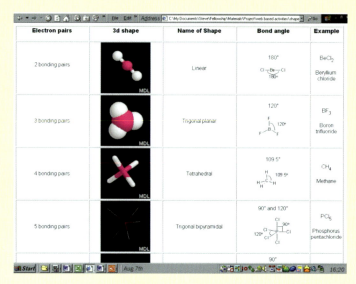

Figure 43

An example of an interactive on-line resource. This is a web page with associated image files (resource 46) that includes molecular representations which require the Chime plug-in to be active. Thanks to Dave Tandy, Solihull Sixth Form College. The molecular images may be interacted with in several ways whilst being viewed on a screen, including changing the display, size or orientation of each representation.

Other molecule file viewers

RasMol was an early molecule file viewer which is still widely used. Several viewers with more advanced features have been developed. More information about these, including free downloads and tutorials, can be found by following some of the project web links or simply typing 'pdb file viewers' into an internet search engine such as: **http://www.google.co.uk**

Website pages: htm or html files

There are many websites supporting chemical education. A large number of these offer no more than text on the screen, leading to the question: 'doesn't a text book do the same and in a more accessible way?'

There is no definitive answer to this, but generally people prefer to make use of information rich text based resources on paper rather than via a computer screen and also tend to use a paper based resource more effectively. There are exceptions to this and opinions do differ.

Take care not to be taken in by well designed layouts, too many websites look good superficially but are simply pages of text of dubious value. Some features to consider when evaluating a website:

- a fast internet connection is essential in order to use some items effectively eg video clips or Java applets;
- interactivity through links to allow individual routes to be mapped;
- interactivity through models to enable exploration and clarification of ideas;
- reviewing of understanding through questioning, but check this offers more than paper question and answer exercises would;
- stimulating use of images eg video;
- high quality text: this doesn't necessarily contradict the points above; if the text content is particularly well written or pertinent then this can be a useful resource, although few examples exist;
- well organised sources of data or reliable references to other sources of information.

If websites are to be accessed live with a class group of students, then careful consideration should be given to the quality of the available internet connection. Is it fast enough and reliable enough to make this a viable activity? If not alternative approaches can be pursued with technical support staff, for instance the possibility of saving the site and its resources to the local computer network so that a live connection isn't required.

Both the **http://www.ChemIT.co.uk** (resource 99) and **http://www.chemsoc.org/learnnet** websites have links to useful and interesting website addresses. Placing them there, rather than providing a list of URLs (Uniform Resource Locators ie the addresses of websites) within these pages, allows them to be checked regularly to ensure they are live. It also means that they can be explored with a single mouse click rather than needing to copy each address into an internet browser from a list on paper.

Still images: gif, jpeg, tiff, bmp files etc

Being able to display appropriate images on a large screen in a chemistry classroom has proved to be a very powerful tool for motivating and informing learners, particularly for those who respond to a strong visual stimulus. Combining text with different types of image in simple animated sequences can make topics clearer and more stimulating.

There are an enormous number of different picture file formats and various sources of copyright free images. These types of files are readily inserted within activities developed using Office®, as described in the Skills guidelines.

The size of an image file needs to be taken into consideration. A balance needs to be judged between having:

- a reasonably clear image when displayed on a screen;
- a prohibitively large file size for the resource that the image is included within, eg if this is to be transferred elsewhere.

There are many factors that contribute to the appearance of an image. These include: resolution, contrast, the proportion of the screen filled, the image file format and the skill of the photographer. Often images that appear reasonable when viewed on a computer monitor are not suitable when projected on to a large screen, particularly if there are high levels of light in the room.

As a rule of thumb, image files above 50 Kb in size often look reasonable when displayed on a screen. Files below this size may be unclear, which can reduce the intended impact and even demotivate students. It is sensible to check what an image looks like on the big screen before using it with learners.

Animated gif files

These are picture files that appear on the screen as cartoons which can be set by their designer to loop once, for a finite number of times, or endlessly. These may be inserted into PowerPoint® presentations or web pages and can be effective illustrations of cyclical processes.

Care needs to be taken when using these animations. The constant repetition can be distracting, particularly so with the light-hearted ones widely used to enliven web pages and readily available from on-line libraries.

The animated gifs illustrated in the previous two figures were both created by Mike Thompson of Truro School. Resource 29 is a PowerPoint® file which includes these and other examples.

There is no user control over the speed at which an animated gif runs and no way to pause the animation. If PowerPoint® is used to create a similar sequence then it is possible to control the flow. The distinction is illustrated in a PowerPoint® file (resource 40) containing both a sequence of PowerPoint® slides and an animated gif of the flow of electrons during electrophilic addition.

Figure 44
Stills from an animated gif illustrating the formation of a covalent bond (resource 29). Animation: Mike Thompson of Truro School
http://www.chem-pics.co.uk

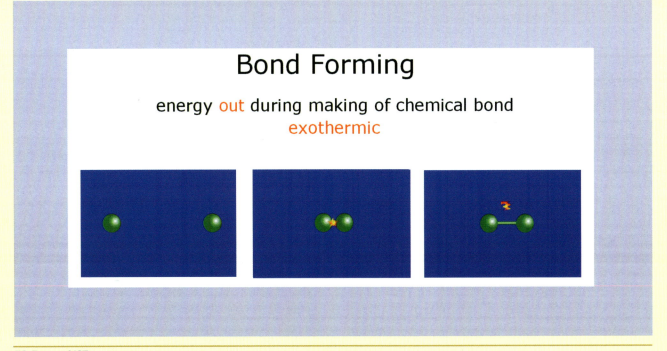

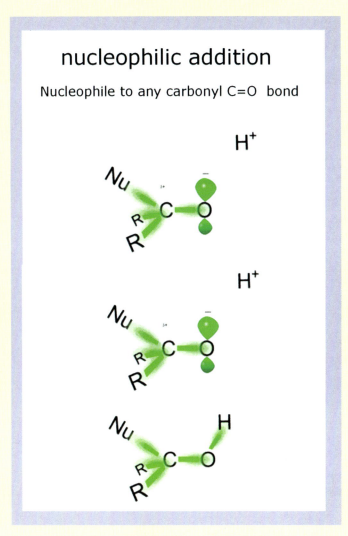

Figure 45

Stills from an animated gif illustrating electron flow during a nucleophilic addition mechanism (resource 29). Animation: Mike Thompson of Truro School

http://www.chem-pics.co.uk

Video clips: avi, mov, mpg files etc

Video clips can be inserted within a PowerPoint® presentation, so that the clip can be played as part of the sequence of events displayed. They may also be inserted within Excel® spreadsheets. The clips can be 'embedded' so that they run within these applications or they can be set to run from a link within media software. The clips can be displayed to run repeatedly or so that individual frames can be frozen and examined more closely. Some examples of chemical reactions are included with the project materials.

Whilst probably not a problem when run from a classroom computer, the large file sizes of video clips becomes a key issue when considering how a resource making use of them is to be distributed. For instance it may be appropriate to use them if the resource is saved on a CDROM but not to a floppy disk. Web based activities using video clips may only be viable viewed via a fast internet connection.

Possible uses of these clips include:

- demonstration of exotic or large scale processes that cannot be carried out in the laboratory;

- shown prior to students carrying out an experiment, perhaps to highlight practical techniques or safety procedures;

- shown immediately after the experiment shown in the video clip has been carried by students out in order to highlight key observations;

- shown some time after the experiment shown in the video clip has been carried out by students to remind them of what they observed, perhaps to act as a visual aid to help in remembering the relevant reaction equations;

- included within review 'electronic worksheets' as a stimulating prompt for questioning.

RS•C

Macromedia animations

Flash/Shockwave (swf files) and Director (dir or dcr files)

Macromedia Flash / Shockwave or Director files are often embedded within web pages to add animation and interactivity. To view these the Flash Player is required for swf files and the Shockwave Player for dir or dcr files. Both are free downloads from **http://www.macromedia.com**.

Flash tends to be used to create relatively simple animations and is very widely used whereas Director is used to create more complex materials, such as games, multi-media or tutorial packages. Many educational resources available on the internet or CDROM make extensive use of these types of file.

If the permission of the owner of the original material has been confirmed, then it is possible to take Macromedia Flash swf or Director dir or dcr files found embedded within web pages and use them within new web based teaching materials. An advantage of doing this is that the individual Macromedia animations can be removed from their original context and integrated into customised electronic lessons or independent learning resources. They might then be displayed so as to fill the whole screen and so be more visible to a class than they might have been within the original context.

Figure 46

*Stills from Director animations of particle movement within the solid (solid.dir), liquid (liquid.dir) and gaseous (gas.dir) states taken, with permission, from the Oxford University Virtual Chemistry Lab's on-line text book. Thanks to Dr Karl Harrison. These are from a collection available for teachers to use from the Oxford University Chemistry Department website at **http://www.chem.ox.ac.uk/vrchemistry/***

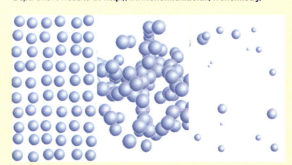

Figure 47

Stills from a Flash animation, Extraction of aluminium.swf (resource 47). Provided by Matt Davis of The Arthur Terry School in Sutton Coldfield.

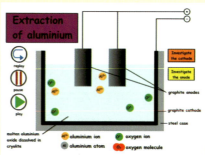

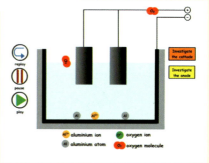

The following sequence launches when 'investigate the cathode' is clicked with the mouse.

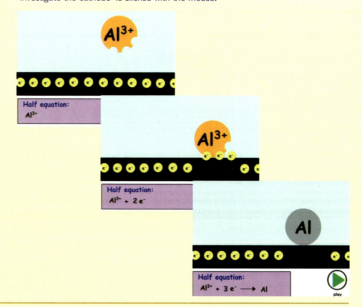

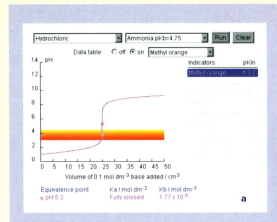

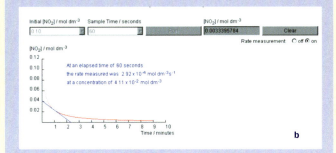

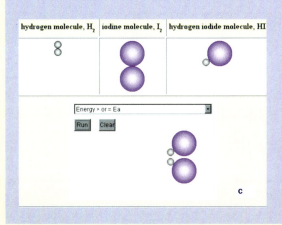

Java Applets

Java can be used to create programs that can be safely downloaded through the Internet and immediately run without fear of viruses or other harm to computers or files. 'Applets' are small Java programs that allow functions such as animations, calculators and simulations to be added to web pages. These are often designed so that the user can interact with and control what appears within the Applet. They are widely used to provide interactive options within on-line tutorials and many websites make use of Java Applets.

Java Applets are not created using JavaScript.

A small collection of Java Applets, each with a webpage of background information are included on the accompanying CDROM. These may be run from the CDROM or saved elsewhere eg as part of a school or college chemistry intranet.

Figure 48

a) screen from an acid base pH titration applet created by Warwick Bailey.

b) screen from a reaction kinetics applet created by Warwick Bailey.

c) screen from a collision theory applet created by Warwick Bailey.

RS•C

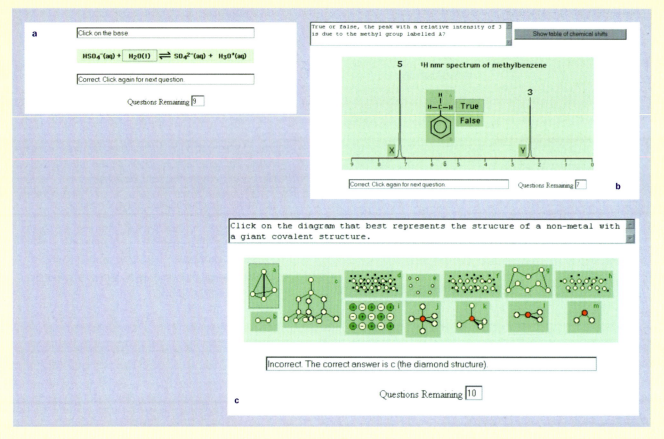

a Click on the base.

$$HSO_4^-(aq) + \boxed{H_2O(l)} \rightleftharpoons SO_4^{2-}(aq) + H_3O^+(aq)$$

Correct. Click again for next question.

Questions Remaining 9

True or false, the peak with a relative intensity of 3 is due to the methyl group labelled A?

Show table of chemical shifts

¹H nmr spectrum of methylbenzene

True
False

Correct. Click again for next question.

Questions Remaining 7

b

c Click on the diagram that best represents the strucure of a non-metal with a giant covalent structure.

Incorrect. The correct answer is c (the diamond structure).

Questions Remaining 10

JavaScript

JavaScript is a programming language often used to add interactive features to web pages. JavaScript is not used to create Java Applets.

Figure 49

Examples of interactive exercises for post-16 students produced using JavaScript within web pages. Thanks to Mike Docker, taken from his website
http://www.mp-docker.demon.co.uk/
a) acids and bases
b) nmr spectroscopy
c) structure and bonding.

Specialist authoring software

There are very many examples of more sophisticated resource authoring software. The two examples here are chosen as reasonable specialist applications for teachers to move on to once they become confident with using the generic Office® software widely available to them.

Hot Potatoes

Hot Potatoes is software from the University of Victoria, British Columbia, Canada. Hot Potatoes is not freeware, but it is free of charge for non-profit educational users who make their pages available on the web. Other users must pay for a license. This is explained more fully on the Half-Baked Software Website, from where the software may be downloaded:
http://web.uvic.ca/hrd/hotpot/

It is used by many teachers to produce a range of interactive 'e-worksheet' resources that run within web pages. These can be created without any need to understand the underlying programming, which makes use of JavaScript. As these resources work within a web environment they are very versatile. As long as the internet 'browser' (software needed to view web pages) used is relatively recent these resources are likely to be compatible with most systems in schools and colleges or on the students' home machines.

Hot Potatoes provides templates and simple tools to create and automatically mark six different types of worksheet:

- interactive multiple-choice;
- complete crossword puzzles;
- type short answer;
- drag and drop jumbled sentence;
- drag and drop matching or ordering;
- gap filling exercises.

There is also a facility for combining several of these to build a single resource.

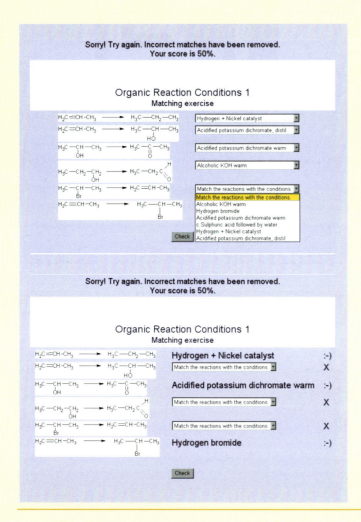

Figure 50 a and b
An example of a 'Hot Potatoes' activity. The interactive multiple choice exercises for post-16 students were produced by Ian Bridgewood of Queen Elizabeth II College, Leicester.
a) Interactive multiple-choice. Drop-down menu choices with a 'check' below that acknowledges the correct answers and leaves the errors to be corrected (resource 41).

Figure 49

c) Interactive crossword puzzle (resource 42).

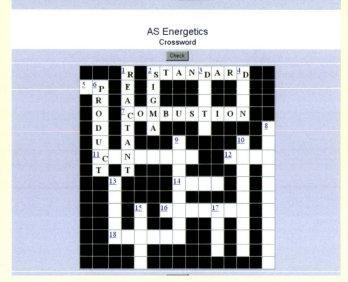

d) Drag and drop jumbled sentence (resource 43)
The sentence has to be exactly as prescribed
in order for the 'check' to declare it correct.

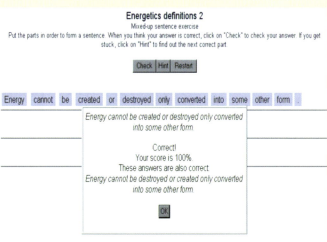

SimChemistry

This is an example of a small software authoring package produced by a chemist. Simchemistry is available as 'freeware' and may be downloaded from **http://www.simchemistry.co.uk/**

It is a tool for creating simulations of particle movements for chemistry teaching and experimentation. It allows a wide range of chemistry phenomena involving particles to be illustrated on-screen and related to the underlying molecular behaviour.

Various features can be built into these simulations, including reactive and non-reactive collisions, strength of inter-particle attractions and the effect of temperature, pressure or volume. Controls can be included that allow the effect of variables on particle behaviour to be explored.

SimChemistry is produced by a chemist for chemistry teachers and as a result has big advantages over trying to use generic software like Office® to create subject specific material. It requires minimal ICT skills and so is accessible to teachers with no interest in becoming computer programmers.

Figure 51

Model of reactive collisions, an examples of a SimChemistry simulation created by the author of SimChemistry, Charlie Wartnaby and used with permission.

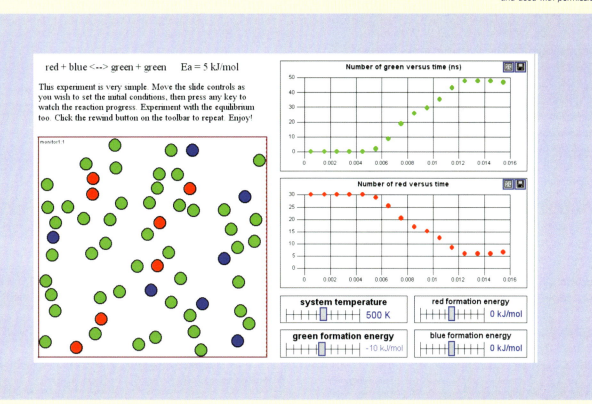

RS•C

Royal Society of Chemistry materials

Many of the previous RSC projects and publications to support chemistry teachers are sources of ideas and materials that lend themselves to an ICT approach and can be incorporated into the classroom electronic lesson presentations and e-worksheets discussed elsewhere within these guidelines. In general, one copy of these materials has been sent to the 'Teacher i/c Chemistry' in UK schools and colleges. Further copies may be ordered via **http://www.chemsoc.org/learnnet**.

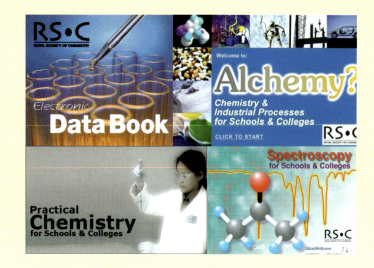

CDROMs

For several years the RSC has been producing CDROMs of resources to support chemistry teaching. It is increasingly making web-based resources available.

These are sources of items to include when creating the type of activities discussed for this project eg image and video clip files.

The complexity of the programming involved means that most of these materials are not customisable by teachers. However, they are presented in small chunks which allows the use of dedicated sections. For more information go to
http://www.chemsoc.org/networks/learnnet/resources.htm

Paper based publications

RSC Teacher Fellow projects

RSC Teacher Fellow projects have covered a diverse range of topics and activities that can be accessed or supported through developing ICT resources.

Contemporary Chemistry for Schools and Colleges (2004) developed by Dr Vanessa Kind, is provided as both pdf and Word files and with associated interactive components, all provided on a CDROM that accompanies the Teachers' Guide. This allows a flexible selection of material on contemporary aspects of chemistry, some of

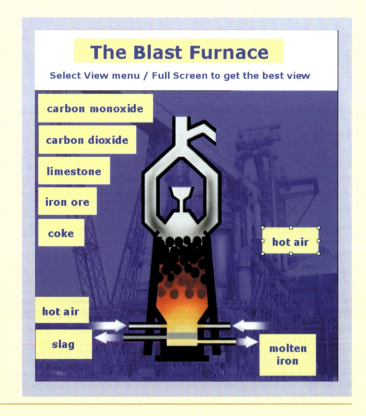

Figure 52

A Word® 'drag and drop' document (resource 44) that uses an image copied from the Alchemy? CDROM.

which can be used to develop other interactive material to that provided on the CDROM should teachers so wish.

Keith Taber's 'Chemical Misconceptions' project (2000–2001) includes a classroom resource pack with a large number of paper based activities. Many of these can be adapted for display on a large screen to the whole class or to provide ideas for e-worksheets.

These documents are available on RSC Learnnet website in both pdf and Word formats at
http://www.chemsoc.org/networks/learnnet/miscon2.htm.

This section of the website also contains other material on misconceptions at
http://www.chemsoc.org/learnnet/misconceptions.htm.

Teachers may also want to use some of the material in 'On the Ball' when teaching particle theory to the early years of secondary school.

Figure 53
Slides from a PowerPoint® presentation (resource 45) based on an activity in the Chemical Misconceptions' Volume II: classroom resources book.

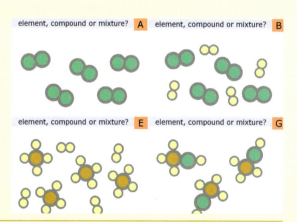

RS•C

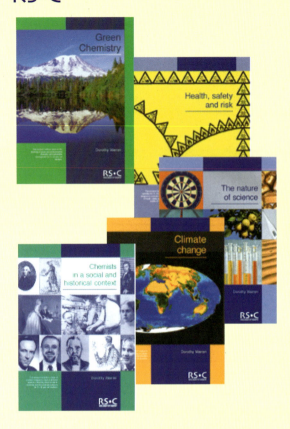

The worksheets in Classic Chemistry Experiments produced by the 1997–1998 Teacher Fellow, Kevin Hutchings, are available on Learnnet in Word® format so teachers can modify or produce differentiated versions of them at **http://www.chemsoc.org/networks/learnnet/classic-exp.htm**.

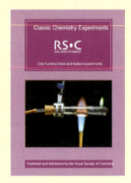

Ted Lister's 'Classic Chemistry Demonstrations' project (1993–1994) includes brief theory notes that can be used as the basis for a large screen presentation to explain the live demonstration.

Dorothy Warren's resources (1999–2000) look at the nature of science and the teaching of investigative skills. These include classroom resource activities and background information that might be incorporated into 'Context' PowerPoint® presentations. This material is made available on the RSC's learnnet website at **http://www.chemsoc.org/networks/learnnet/ ideas-evidence.htm**.

Video clips of several of these demonstrations can be found at the Oxford University Chemistry Department website within a 'Chemistry Film Studio': **http://www.chem.ox.ac.uk/vrchemistry/FilmStudio/fshome.htm**

Links to these could support a follow up activity after having seen the demonstration. Note that a fast internet connection is necessary to view video clips on-line.

There have been several publications that can act as the basis for 'context' PowerPoint® presentations. Ideally these will be supported by images, including ones found by searching the internet (taking copyright considerations into account) as described in the 'Skills' sections.

Web based

The RSC also produces materials that are only available on websites. These include the Chemistry Now series, Green Chemistry for Schools and Colleges, and the JESEI (Joint Earth Sciences Education Initiative) material.

Chemistry Now

Chemistry Now is a series of leaflets designed to present modern aspects of chemistry in a way that is accessible to school students and directly usable by teachers. There are four leaflets currently available:

- Chemistry and sport
- Chemistry of the atmosphere
- Computational chemistry
- Combinatorial chemistry

Each leaflet consists of four pages of information interspersed with questions to test the student's understanding of what they are reading, to help them to link what they have read to the chemistry they already know and to help them to understand the text. The leaflets may be used to support existing work schemes, to develop

comprehension skills or as meaningful exercises for use in case of teacher absences (planned or unplanned). The answers leaflet is primarily for the use of teachers and contains some background information and answers to all of the questions. These are to be found at **http://www.chemsoc.org/networks/learnnet/ chemnow_ans.htm**

Green Chemistry

Green Chemistry for Schools and Colleges is the result of a collaboration between the Royal Society of Chemistry, the American Chemical Society and the Gesellschaft Deutscher Chemie. Three versions are available, customised for each country, and with content appropriate to that country. This can be found at **http://www.chemsoc.org/networks/learnnet/ green/index.htm**

In the United Kingdom it can be used to support post-16 and undergraduate chemistry courses and gives up-to-date contexts for the study of topics such as yield, areas of organic chemistry, and intermolecular forces.

Additional background reading is provided, as are references for teachers and students to explore Green Chemistry in more detail.

The resource builds on the RSC paper resource Green Chemistry by Dorothy Warren produced for 14–16 year old students and their teachers.

Joint Earth Science Education Initiative (JESEI)

The JESEI website aids Chemistry, Biology and Physics specialists with their teaching of Earth Science by providing material within each science specialism. Resources include worksheets, weblinks, and teachers notes. The worksheets and notes enable teachers to build up their background knowledge and also outline interactive practical activities that can bring the teaching of earth science to life.

RS•C

The website is a joint initiative developed by the RSC in conjunction with the Institute of Biology (IOB), the Institute of Physics (IOP), the Earth Science Teachers' Association (ESTA), the Royal Society (RS), The Association for Science Education (ASE), the Geological Society and the UK Offshore Operators Association (UKOOA). The website is at **http://www.chemsoc.org/networks/learnnet/jesei**.

Datalogging

Datalogging is often presented as a means of using ICT during science lessons, particularly pre-16. Caution is needed that this doesn't become the main purpose, rather than using it to provide opportunities to support teaching and learning.

The comments below focus on ways in which teachers in schools and colleges in the UK are using datalogging to enhance experimental activities, rather than using the technology as an end in itself.

Equipment

Sensors are used to monitor variable quantities.

Most commonly used:
temperature (including corrosion resistant probes), light, pH

Also available: pressure, oxygen, conductivity, humidity, motion

Specialist:

- many balances have data outputs to computers. These require the purchase of a connecting lead and software from the balance manufacturer;

- colorimeters often have the facility to connect the output to a computer, again requiring software to interpret the data.

There are a large number of different datalogging systems available. The following is an overview and some systems offer different approaches and features. When choosing a system it is sensible to spend time looking at the different options available.

One or more sensors are connected to a datalogger, of which there are several types available. The datalogger collects and often stores data detected by the sensor. The datalogger then passes the information to a computer (either as it is collected or downloaded later) where specialist software can be used to display and analyse the data. Usually the software is also used to control the timing and value range of the sensing, some dataloggers allow this without the computer connection. Datalogging software usually allows data to be exported to generic data handling applications such as Microsoft Excel®. Summarising:

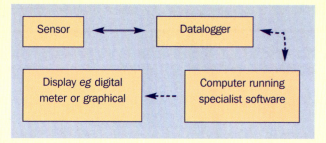

Advantages

By automating the collection and display of experimental data, datalogging can allow the emphasis of a particular practical session to be on the analysis and evaluation of results. This can avoid learners being distracted by practical procedures.

Some ways in which datalogging and associated software can enhance traditional practical procedures include:

- changes that are too slow or fast to be followed by other procedures;

- tedious procedures can be followed automatically allowing attention to be focused on analysis of results eg obtaining cooling curves or pH curves;

- experiments can be repeated quickly eg to allow the effect of changing conditions to be investigated or to demonstrate the effect of experimental error on the outcome of the same experiment;

- several variables may be monitored and displayed simultaneously *eg* temperature and pH for a neutralisation reaction;
- data can be made visible to the whole class on a large screen as it is collected *eg* as a digital meter reading;
- for project work, as a way of speeding up initial trials before students decide on suitable parameters for their experimental work;
- teachers enthusiastic about using datalogging have said they find teaching these lessons enjoyable and that it allows more time for extra-curricula work.

Issues

- The relatively limited number of links to the curriculum where datalogging is relevant, especially for post-16 work.
- Datalogging equipment is significantly expensive for most institutions, particularly if class sets are required.
- The complexity of setting up and managing the sensor – datalogger – computer – software interconnections effectively within lesson time.
- Teachers often have difficulty finding the time to familiarise themselves with the technology in order to feel comfortable using it with their students.
- Although this technology has the potential to allow learners to focus on experimental analysis and evaluation, careful classroom management is needed to prevent the datalogging technology distracting attention from thinking about the science.

ICT Skills for teachers of chemistry

Introduction

'It would appear that we have reached the limits of what it is possible to achieve with computer technology, although one should be careful with such statements, as they tend to sound pretty silly in five years.'

John Von Neumann (*ca.* 1949)

When it comes to ICT, teachers of chemistry inevitably have a wide range of skills, interests and experiences. These guidelines are mainly intended for teachers with basic skills (eg familiarity with basic word processing), to enable them to create the type of materials described in the Ideas sections and illustrated by resources available from the project website at **http://www.ChemIT.co.uk** and also at **http://www.chemsoc.org/learnnet/chemIT**.

These notes include hints and suggested approaches collected from various sources, including comments from participants at numerous training events. It is hoped that even more experienced users of ICT may still find something novel or that stimulates a new approach by sampling these pages.

The notes have been written so that they may be followed in sequence as a tutorial, but also assuming that the contents pages will be used to explore particular skills in isolation when it is appropriate to do so. They are not intended to be a comprehensive handbook, rather they are a compilation of sufficient skills to get started and gain confidence, together with useful tips and shortcuts, many of particular relevance to teachers of chemistry.

It has to be stressed that there is no definitive approach to designing and creating these types of materials. People have widely differing ideas about what constitutes an attractive design. There are often several ways to achieve similar outcomes when using generic software such as Office® applications. These guidelines should be approached open-mindedly, since experience leads people to evolve their own approaches to suit their own tastes and needs.

A significant difficulty with notes such as these is how quickly resources available to teachers change. During the time span of the Teacher Fellow project the majority of teachers were using Microsoft Office® 2000 applications and the detailed instructions that follow reflect this.

On the whole, more recent versions of the software have broadly similar screen appearances and organisation of menus. The XP (2002) version of PowerPoint® does have a different screen layout and a few additional features compared to previous versions, and so a separate introductory section is included for this.

An approach many 'User Manuals' follow is to include labelled computer screen-shots illustrating sequences of steps. Although these can make instructions clear, their use here has been kept to a minimum to prolong 'shelf-life' as screen layouts change with software updates, and also to keep the document to a reasonable length.

Toolbars

Go to the View menu and select Toolbar in order to display any toolbar.

Hover the mouse over any toolbar button to get description of what it does.

For example, the Standard Toolbar:

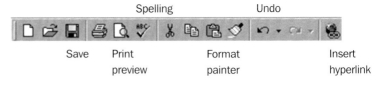

Re-sizing objects (pictures, drawings, text boxes etc)

Resize any object by pointing at it with the mouse, left-clicking once to select it then left-click drag a white sizing handle

Note that dragging a corner sizing handle maintains the height:width ratio, whereas dragging a central handle will distort the ratio.

Drawing toolbar

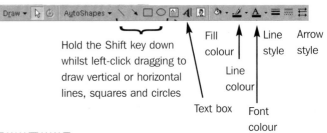

Hold the Shift key down whilst left-click dragging to draw vertical or horizontal lines, squares and circles

Fill colour

Line colour

Line style

Arrow style

Text box

Font colour

Text Boxes

To edit the contents of a text box, left-click within it and type.

To change the format of the text box itself (as opposed to its contents) display its outline by left-clicking within it and then:

select the text box by left-clicking once within its border

or resize a text box by left-click dragging a white sizing handle.

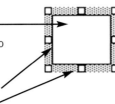

Copying and Pasting or Duplicating

These three are so widely used and can save so much time that it is worth emphasising how they can be done.

To copy any item to the Clipboard and paste it elsewhere.

* use the mouse to select the item
* select copy (keyboard Ctrl+C, right click mouse menu or Edit menu)
* place the mouse cursor where you want to paste
* select paste (keyboard Ctrl+V, right click mouse menu or Edit menu).

With PowerPoint® and Word® Drawing there is the very useful additional facility to duplicate items:

* use the mouse to select the item
* use the keyboard keys Ctrl+D

a duplicate should appear close to the original.

Alternatively, using a combination of mouse and keyboard:

* highlight / select what is to be copied or duplicated,
* whilst pressing the control key, point at what is selected and left-click drag and drop it where a copy / duplicate is required and then release the control key.
* pressing both control and shift keys together whilst drag and dropping forces the copy / duplicate to remain in either vertical or horizontal alignment with the initial object.

Screen Prints

Pressing the Print Screen key on the keyboard (how this key is labeled varies but it is usually found above the Insert key) copies all of the image on the screen to the clipboard so that it can pasted into any file.

The resulting image can be edited and formatted using the Picture Toolbar.

To copy an image of the active window, *ie* the one with a coloured bar at the top, press the Alt and Print Screen keys together.

Microsoft Powerpoint®

Introduction

Many teachers use PowerPoint® as an ICT 'backbone' to organise the structure of their lessons *ie* not only to provide animated slides of information but also as a way of linking, often seamlessly, to other software applications such as video clips or simulations. These guidelines reflect this approach, which means that as well as suggesting skills that might be useful to teachers of chemistry using PowerPoint®, these pages cover several other ICT applications. Consequently, many of the suggestions within the 'PowerPoint®' sections are also applicable to resources created using other applications such as Word® and Excel®.

These suggestions make no reference to any of the templates or design formats provided by Microsoft, but instead start with a 'blank page'. Many teachers report that this allows them more creativity and also leads to a clearer understanding of how sequences for teaching and learning might be best constructed. However, this philosophy is not universal and significant numbers of teachers (and many professional 'presenters') make extensive use of 'built-in' features, *eg* for background designs, formats and the more dramatic animations.

The first section, *Creating a new PowerPoint® slide*, consists of detailed instructions intended for people with minimal experience of using PowerPoint®.

Those with some prior experience may prefer to start with the second section, *Creating a sequence of animated slides*, which includes guidelines for drawing chemical formulae and structures by making use of the generic Draw software available in several Microsoft applications. This section also suggests various ways of integrating other ICT applications within a PowerPoint® resource *eg* inserting images, including video clips and Macromedia Flash® animations.

The introductory section is repeated for PowerPoint XP®, since this has a significantly different screen layout compared with previous versions. XP also offers a few additional features, and a brief outline of some of these is included.

RS•C

Presenting or teaching

Microsoft frequently refer to PowerPoint® files as 'slide presentations'. Inevitably this description is used when discussing PowerPoint® based resources created for teaching and learning. Several teachers have objected to this terminology on the reasonable basis that they do not 'present' lessons to students, rather their teaching is a highly interactive process.

It is important to keep this in mind when designing these resources eg to build in a 'what happens next?' element through using prompts and revealing information within stepped sequences or through making exploratory choice possible through the use of links to other slides or files. Good teaching does not correspond to good presentation alone.

The experience can be made dynamic and interactive if the teacher constantly challenges learners to anticipate 'what is going to happen next', both verbally and through getting them to write things down. Strong images can be used to initiate discussion. Building up complex ideas on the screen step-wise can promote understanding. A better term than 'presentations' might be 'stimulations'.

Ensuring attention

As with all teaching techniques, occasionally some members of the class become inattentive or distracted. Turning the screen black during a PowerPoint® slide show (simply by pressing the B key), can help the teacher to retain attention. Turning the screen white (simply by pressing the W key) allows the teacher to turn attention to the board, an OHP transparency, a model etc.

Design considerations

When creating 'slide shows' it is useful to bear in mind a few essential tips.

- Too many fancy effects and animations can detract the audience's attention away from the content and may also prove annoying.

- Keep each slide fairly simple, don't try to cram in too much information. However, using this software to gradually reveal the components of a complex idea or image can be very effective.

- Consider using the same background colour, font style and size throughout for the same reasons as above, although this does depend on your intended audience. For instance, a younger audience may respond better to a greater variety of styles. An advantage of using PowerPoint® is that the teacher has control over this and so is able to shape the materials to the needs of a particular group.

Creating a new PowerPoint® slide

If PowerPoint XP® is being used turn to the section on page 99 describing this.

Note that for most of these instructions there are alternative ways of achieving the same result. For instance, many actions that use the toolbar menus can also be carried out through the mouse right-click menu. Colleagues (and students) are good sources of alternatives, especially the 'shortcuts' that save time. Individuals will find the approaches most appropriate to their needs through exploring the alternatives.

Opening a new PowerPoint® file and screen layouts

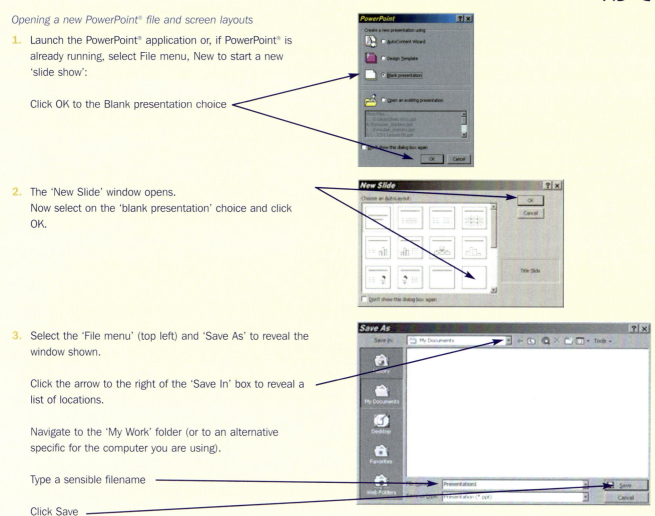

1. Launch the PowerPoint® application or, if PowerPoint® is already running, select File menu, New to start a new 'slide show':

   Click OK to the Blank presentation choice

2. The 'New Slide' window opens.
   Now select on the 'blank presentation' choice and click OK.

3. Select the 'File menu' (top left) and 'Save As' to reveal the window shown.

   Click the arrow to the right of the 'Save In' box to reveal a list of locations.

   Navigate to the 'My Work' folder (or to an alternative specific for the computer you are using).

   Type a sensible filename

   Click Save

The screen is probably displaying Normal view in PowerPoint®.

If the screen does not look like the screenshot below then select the View menu and choose Normal.

Note the two parts of the screen:

On the left ————
Provides a summary of
the whole presentation.

On the right ————
The current slide from the new presentation, which can be edited in this view.

Click this small button (hovering the mouse over it reveals the label 'slide show') to run the presentation for the slide displayed, as opposed to the whole slideshow from the beginning.

The following steps are a guide to producing a title, heading, reaction reactants and products which appear on the screen in sequence controlled by the mouse or keyboard.

Note that should you make a mistake at any point, clicking the Undo button (backward curly arrow) will restore the previous setting.

A PowerPoint® file named Sample PowerPoint.ppt (resource 48) contains slides created following these introductory guidelines for comparison. This is available to download from the **http://www.ChemIT.co.uk** website and on the CDROM.

Note that these instructions are intended to support the development of confidence in working with PowerPoint®. There are many different ways to achieve the same results and it is not suggested that the methods here are any better than the alternatives. Individuals will find the ways that best suit through practice, exploration and picking up tips from others.

Adding a title

Click immediately to the left of the slide picture.

1. Type some text and a title text box will automatically appear on the slide *eg* Chemical Presentation.

2. Click the edge of the title text box once and its border will be highlighted.
    This means that the text box itself has been selected and the format of its contents may be changed.
    Try clicking once inside the text box and compare the difference. This selects the text inside the box and allows this to be edited.

3. With the edge of the text box selected, change the font and font size by making choices from the menu at the top of the screen

4. Open the Drawing toolbar by selecting View Menu, Toolbars, Drawing. This will probably appear along the bottom of the screen.

5. With the title text box selected (click its edge remember), use the Drawing toolbar button to change the Font Colour.

> Note that 'hovering' the mouse over a button for a few seconds displays information about what the button does.

6. Save your work (File menu, Save).

Adding a new text box

1. Click the Text Box button in the Drawing toolbar and drag a new text box below the title by holding the left mouse button and dragging the box across the screen. Type some new text in *eg* Industrial preparation of hydrogen gas from methane. Format the Font, Size, Colour for this new text box.

2. Click the edge of the text box and then drag it to approximately the required position on the slide.

Positioning objects on a slide

There are various other ways to position objects precisely. The object must first be selected. The arrow keys allow fine control and using an arrow key whilst pressing the control key down allows 'nudging' of the object.

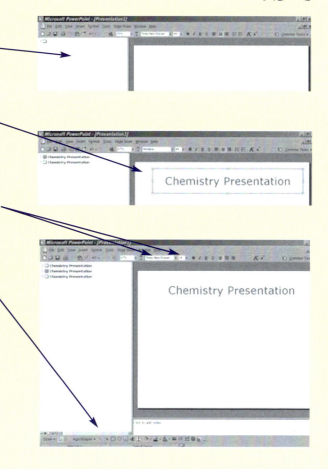

To align several objects simultaneously, either with each other or the slide:

1. First select them all by either clicking the Select Objects button in the Drawing toolbar and dragging the mouse over the objects to be selected,
    or holding the shift key down and left mouse clicking each object in turn.

2. Then use the Align feature from the Draw menu of the Drawing toolbar.

Animating text and writing a reaction equation

The second text box can be given an animation effect:

1. Click the edge of the new text box to select it and then click the Slide Show menu and select Custom Animation to open a window:

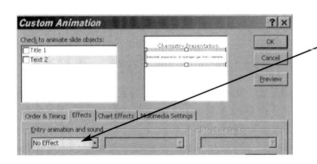

Select Entry animation from the list *eg* Appear and OK.

2. Save your work.

3. View the slide show created so far by pressing the F5 key (this is one of several ways to start a PowerPoint® show). There are several ways to move through the animations, these include:

    *   left click the mouse; press the space bar; press any of the return (enter) or right arrow or down arrow key.

    *   pressing any of the backspace or left arrow or up arrow keys reverses the animation sequence.

Duplicating text boxes

Now add a third text box. This can be done as above or alternatively the second one may be duplicated to make a third, which has the advantage of copying all formats, including animation effects, and therefore saving time and ensuring consistency of design. To duplicate a PowerPoint® text box (and other objects):

1. Click the edge of the second text box (*ie* the one animated in the previous step) to select it, hold the Ctrl key down and then, without releasing Ctrl, press the D key. This should duplicate the title text box.

2. Use the mouse to position the duplicate on the page (click the edge then 'drag and drop', then if required use the precise positioning described above), re-set the font size and colour as required.

The following instructions may be applied to type any reaction equation and work when creating Word® documents too *eg* to produce the equation:

$$CH_4(g) + H_2O(l) \rightarrow CO(g) + 3H_2(g)$$

3. Type CH then, before typing the '4', hold the Ctrl key down and, without releasing Ctrl, press the = key. This sets the following characters as subscript. Repeat this (*ie* Ctrl with =) to turn subscript off.

   NB To create a superscript hold the Ctrl key down and then, without releasing Ctrl, press both the Shift and = keys. To turn superscript off repeat this (*ie* Ctrl with Shift and =).

   Finish off the reactants by typing '+ $H_2O(l)$'. Position the text box on the slide by clicking the edge and using 'drag and drop'. Save the work.

   To create a reaction arrow try typing – –> (dash dash greater than) after the reactants, which should automatically become →. If this approach doesn't work, use the arrow button on the Drawing toolbar, select the arrow and give it a custom animation of its own (see above).

4. Duplicate the reactant text box and replace the reactants with the products:
   Move the product box to an appropriate position on the slide, note ways of aligning the text described above.

   Save the work and then view the animated presentation (F5 key).

Creating a sequence of animated slides

This section assumes some familiarity with using PowerPoint® and is therefore not as explicit as the previous introductory section, although there is some overlap. It is written with reference to PowerPoint 2000®. Although there are some differences in screen

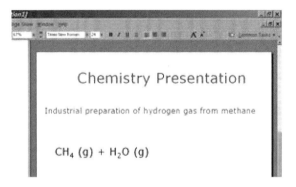

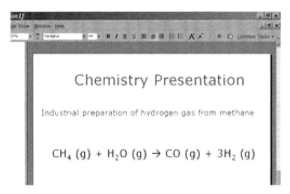

layouts with XP and later versions, these guidelines should still be applicable.

The following pages describe how different slides might be created so that various objects (eg text, images or graphs) appear in a pre-set sequence which is controlled using the keyboard or mouse.

The theme suggested here is to create a slide sequence illustrating various aspects of chemical reactions. A sample PowerPoint® with suggestions for ways to write reaction equations, illustrate these with still images, video clips, animated simulations and drawings of structures are included.

A PowerPoint® presentation containing photographs to use when developing these skills plus a finished version of the tutorial below for reference is available to download from
http://www.ChemIT.co.uk
The file is resource Sample PowerPoint®.ppt (resource 48).

> Note that should you make a mistake at any point, clicking the Undo button (backward curly arrow) will restore the previous setting.

Apply formats to every slide: the Slide Master

The Slide Master is very useful since whatever is on it will appear on every slide in the presentation.

1. Select View menu / Master / Slide Master.
   Re-format the 'Master title style' by first clicking its edge to select it and then choosing a different Font (eg Verdana), Font Size (eg 36) from the Formatting toolbar or from Format menu / Font. This style will now be repeated whenever a title is typed next to the icon for a slide in the left-hand 'summary' window of the edit view.

2. If the Drawing toolbar isn't displayed, select View menu / Toolbars / Drawing. The toolbar ususually appears at the bottom of the screen. Use this menu to change the colour of the Master title text box and it's font using the appropriate buttons.

3. Right click the edge of the slide, taking care not to click inside any of the text boxes, select Background from the menu and choose a colour.

4. To place a slide number on each slide automatically use View menu, Header and Footer, Slide tab, check the Slide number box and click Apply to all. The format can be edited in the Number Area box in the bottom right corner

5. Click to close the Slide Master View.

Apply formats to an individual slide

1. In the left-hand Outline window, click to the right of the slide icon and type a title eg 'Chemical Reactions'. This should automatically appear on the slide itself. Adjust the text box size to suit by click-dragging the sizing squares at the edge of the title text box.

2. Click the Text Box button in the Drawing menu and drag a new text box below the title. Click the edge of the new text box, drag it to where you want it on the screen, highlight all of its text and type some new text eg 'Acid and metal'. Change the text box Fill colour, the Font colour and Font Size of its text. Re-size it.

3. Select the two text boxes by holding down the shift key and clicking the edge of each in turn, then use Draw menu / Align or Distribute / eg Left

Apply Custom Animations

To make the new text box created in the previous section appear on the slide as part of a sequence of events give it a 'Custom Animation':

1. Click the edge of the sub-heading text box to select it, right click the mouse and select Custom Animation (not the only way to access this menu). Click the Add Effect button find Appear from the Entrance list.

2. Save your presentation.

3. To view what you've done so far, click the tiny Slide Show button (towards bottom left of the screen, with an image looking like a screen for a projector, see page 78 for diagram). Alternatively, select the Slide Show menu and

choose View Show, which starts at the beginning of the presentation. Use the forward arrow key, left mouse click or press the space bar to advance the animation sequence.

Typing reaction equations (subscripts, superscripts, arrows)

1. Duplicate the 'Acid and metal' text box by pressing the control and D keys together. Note that duplicating automatically copies the appear animation and adds it next in the sequence. Remove the fill colour (select No Fill), change the font size and colour.

2. In the new text box, type the chemical formulae for reactants
   eg $Mg(s) + H_2SO_4(aq) \rightarrow$ can be typed as follows:

subscripts	control and = keys together to switch on and off
superscripts	control, shift and = keys together to switch on and off
$\rightarrow$ key	hyphen key twice (– –) then > (shift) then space

3. Duplicate this text box and type products:
   $MgSO_4(s) + H_2(g)$
   Position this appropriately: Select the two equation text boxes by holding down the shift key and clicking the edge of each in turn, then use Draw menu / Align or Distribute / Middle.

4. Save your work.

Inserting a photograph

Without closing the presentation, open the Sample PowerPoint.ppt file (resource 48).

1. Open the ppt file and select the photo on the third slide by pointing at it with the mouse and left clicking, then press both the control and C keys together (or use any other of the numerous ways to copy).

2. Move back to your Reaction Equations presentation and paste by pressing both the control and V keys together (or other means).

The photo doesn't fit on the slide.

3. Right click the photo and select Show Picture Toolbar. Select the Crop tool and crop out the pipette from the top of the picture. Re-size and move the remainder as appropriate.

4. Select the photo again. As above, apply an animation: right click the mouse, select Custom Animation, click the Add Effect button find Appear from the Entrance list.

5. Note the Order & Timing tab allows control over the order in which objects appear on the screen. Check that the animation order is as required.

Any image that you see in any PowerPoint® presentation can be copied and edited in this way, but read the advice below regarding the size of these files.

Duplicating an existing slide

A new slide may be created by placing the cursor in the left-hand Outline view window at the end of the title of the first slide and pressing the enter (return) key. However, duplicating an existing slide containing formats and animations that follow a useful sequence is often preferable as it can save time and produce consistent designs.

1. Select the 'Acid and metal' slide created above by clicking its icon in the left hand Outline view window.

2. Duplicate it (hold the control key and press the D key).

3. Edit the duplicate to produce a new slide for the reaction between acids and carbonates. Type 'Acid and carbonate' in the main title text box.

   Useful time saver: select the 'Acid and metal' sub-heading (click edge of its box), click the Format Painter button (in the Standard toolbar) to copy its formats and then click the title text box to convert this to the same format.

4. Delete the old sub-heading and the Mg reaction photo.

5. Edit the reactant and product boxes, using hydrochloric acid and calcium carbonate as the example.

6. Find a relevant image of this reaction eg on a CDROM or

RS•C

by searching the internet (follow the section below about searching for and inserting Images) and if copyright permissions allow, paste this to the slide.

7. Save the work and run the slide show (F5).

The results may be compared with the Sample PowerPoint.ppt file (resource 48).

Navigation

Hyperlinks add another dimension to PowerPoint® files, allowing the user to break out of a linear sequence of events and move to another slide within the same file or to a different ICT application (including a live website), but without losing the path back to the original link.

Any object on a PowerPoint® slide can be made a hyperlink eg text within a text box, the text box itself, any shape drawn on the page, simulated buttons available from the Drawing toolbar etc.

There are several ways to create a hyperlink. One method is:

1. Select the object or text.

2. Right click the mouse and select Action Settings (this appears automatically if you use an AutoShapes button).

3. Using either the Mouse Click or Mouse Over tab, click Hyperlink to and select the target for the link from the options in the drop-down menu.

Two particularly useful navigation hyperlinks include are:

1. To the first slide, especially if this has hyperlinks to the rest of the file.
   There is a Home Action Button in the AutoShapes menu of the PowerPoint® Drawing toolbar which automatically points to first slide.

2. To the previously visited slide. This is very useful for making sure the user finds it easy to follow a sequence, but has the option of clicking away from this without loosing the main thread of the activity. To create such a link select Last slide viewed from the list of link targets in the Action Settings window.

Note that hyperlinks may be created within Word® or Excel® in

a similar fashion to that described here. For instance in Word®, select some text, right click the mouse, choose Hyperlink from the menu and then set the target place or file for the link.

Using the master slide

Anything added to the master slide will appear on all the slides in a presentation. This is useful for adding navigation buttons for long presentations eg Home and Last slide viewed buttons on every page:

1. Select the View menu / Master / Slide Master.

2. Select Drawing toolbar / AutoShapes / Action buttons and click the House icon. Note left click–drag the mouse on the screen to draw the button. A menu automatically appears with the default hyperlink to the First Slide, so OK this.

3. Move the button to the top right corner of the screen,

4. Close the Slide Master to return to the Normal (slide editing) view by clicking the word 'Close' in the small toolbar that should have appeared. Alternatively select View menu, Normal.

5. Run the presentation and see how clicking the button when looking at the second or third slides opens the first slide.

Bookmarks

This is an alternative name for a hyperlink to a target within the same file. If a hyperlink is created by using the mouse to right click an object and selecting Hyperlink from the menu, Bookmark is one of the options. Clicking this reveals a list of all the slides and that the required target may be selected from this.

Hiding slides

Sometimes it can be useful to have hidden slides that are only accessible when the file is run as a show by using a hyperlink to them. To hide the slide currently selected for editing ie the one in view, select Slide Show menu / Hide Slide.

The section suggesting how to create an 'interactive diagram' e-worksheet makes use of several of the features described above.

Automating Presentations

A slide show can be set up to run automatically from start to finish using predetermined timings eg to have it running as a loop for an Open Evening. To do this select Slide Show from the top toolbar and choose the Set up Show option. Here it is possible to set selected slides or the whole slideshow to run through its sequence automatically.

Selecting Slide Transition from the Slide Show menu will also allow control of the way in which each slide in the show is opened and closed.

If a show automatically advances it is vital to get the timings right. Too long a gap between effects means an audience will get bored and too short a gap will mean they will not have time to assimilate information.

More elaborate sequencing is possible. For instance, animations can be set to run and 'loop' (repeat) automatically until the 'Esc' key is pressed.

To summarise, these effects can be created using combinations of settings in the Custom Animation window and the Set Up Show and Slide Transition options in the Slide Show menu.

Printing

Caution: if the Print button is clicked then each separate slide will be printed automatically out on a separate piece of paper.

A handout is usually more useful for printing. This places several slides on one printed page. Before printing set up the layout of the handout: select View menu then Master then Handout Master. This view allows the number of slide images per page and various headers and footers to be set. Note that it is likely that the default header or footer font size and styles will need to be changed.

Image Files

Useful pictures or images are readily found from a range of sources eg other teachers' PowerPoint® presentations, the internet or commercial CDROMs. It is usually a straightforward process to copy and save these elsewhere so that they may be used within a new resource, subject to copyright restrictions.

Once an image is added to a PowerPoint® slide it may be given animation properties so that it appears within an ordered sequence. The order animations appear in can be changed in the Custom Animation window by selecting them and using the 'Re-Order' arrows at the bottom of the Order & Timing tab.

Copyright issues

It is essential to be aware that copyright restrictions may forbid the re-use of images (and other materials). In fact for many materials re-use is allowed for non-profit educational purposes as long as a clear reference to the source is quoted. If a statement that this is permitted is not explicit then it is best to assume that the material should not be used. Experience suggests that if sources are contacted, especially if they are non-commercial, they will readily give permission for educational use.

Inserting or pasting images

Previously guidelines were given for copying an image from one PowerPoint® slide to another. Pictures are often found as separate entities (with filenames such as jpg, gif etc) and assuming copyright permission to do so has been given, these may inserted into a file:

1. Select Insert menu / Picture / From File.

2. Navigate to the folder where the images are saved.

3. Select the required image and click OK.

Inserting images in this way often results in a smaller final file size (particularly for the commonly used jpg format) than the less involved process of copying and pasting the image from where it is displayed on a page or slide.

It is sensible to save a copy of the original file for any image used within a resource. There are special considerations when saving or copying images from the internet, assuming copyright permission is available for this. Take particular note of the use of the Paste Special feature as described in the following section.

Taking images from the World Wide Web

The web is a great source of exciting images, video clips, animations etc. However, unless a website has a notice that says otherwise, its contents are not copyright free. Without such a notice it must be assumed that there are no rights to remove, and most certainly, none to distribute any materials found on the web. If something useful is found then the owner of the website should be approached for permission to remove this from the context within which they are found.

These instructions are for obtaining images for which permission to re-use them has been granted:

1. Open Microsoft Internet Explorer and navigate to a search engine with an image search facility *eg* **http://www.google.co.uk** or **http://uk.altavista.com/** and in either case select the Image search tab. Type some search text. Follow the link from the 'thumbnail' to the source for a suitable image. Check that it is permissible to copy the image. Do not save or copy the image yet.

2. First copy the URL (address) of the website the image is on:

   • If you are copying the address (or any other text) from somewhere on a web page then use the mouse to select (*ie* 'highlight') the address, copy it (Ctrl + C

keys), move back to the where the image is to be placed (eg click tab at bottom of the screen or use the Alt + Tab keys together), select Edit menu then Paste Special (*ie* not Paste alone) and click Unformatted Text. Using Paste Special ensures that no web formats are copied.

   • If you are copying a URL from the address box of a browser toolbar then simply paste (use Ctrl + V keys) as this is text only, without formats.

It is very useful to copy the URL from which the image was obtained so that:

• the source can be found again;

• a reference to its source can be quoted in your presentation, which is often all the owner of an image will ask for in return for granting permission to use it.

It is usually possible to simply copy and 'paste special' the image to wherever it is required. However, for the reasons set out above, it is preferable to save the image file to a folder and then insert the image from there:

1. Point the mouse at the image and right click the mouse to reveal a menu.

2. Select Save Picture As and navigate to a folder where the image file can be stored. Often it is sensible to do this within (or near) the same folder as the resource file that will use the image.

3. The image may now be inserted within the file using the Insert menu, Picture then From File.

Paste Special

If the time saving (though ultimately less useful) 'copy and paste' approach is preferred then it is very important to use Paste Special rather than paste alone.

1. Find the image to be copied, click it with the right mouse button and select Copy from the menu.

2. Open the target file for the image and, it is advisable to

get into the habit of doing this: select Edit menu, Paste Special (*ie* not Paste) and choose Device Independent Bitmap. This ensures that no hidden internet formats are copied into the new file.

3. Once the image has been pasted it can be re-sized, cropped and, if on a PowerPoint® slide, custom animations may be added as appropriate.

Image file sizes

Some images can be large files and if the resource containing them is to be copied elsewhere, a large file size may be inconvenient *eg* transfer may be slow or the file may be too big for the chosen transfer medium.

The procedure used to include an image within an application can have a significant effect on the size of the file created. Using Paste Special and then Device Independent Bitmap is very straightforward and quick. However, with some image formats this method results in files that are significantly larger than necessary, as much as five times more.

Two widely used image formats are jpg and gif. When jpg images are copied using paste special much larger files are produced than when the image is first saved to a folder (right click the image and select save rather than copy) and then inserted into an application (*eg* PowerPoint®) using Insert menu, Picture, From File. The same is not true for a gif image, where the size of the file containing the image is the same whichever method (paste special or insert picture) are used.

Projecting images to a screen

It is quite common for pictures to look clear and sharp on a computer screen but to appear 'washed out' when projected on a larger screen. The Picture toolbar has Contrast and Brightness controls that may often improve the appearance of an image on the large screen.

Internet sources of photographs

A directory of internet image search engines can be found at: **http://www.search-engine-index.co.uk/Images_Search/** (June 2004)

The originators of the web links below have all given permission for their photo files to be used by teachers subject to the following conditions:

Teachers may copy any of these images into any medium in order to create not-for-profit educational resources. They may not include them in any material for commercial gain without prior permission.

Any copyright statement or URL included as part of an image should not be cropped out unless both of these are clearly stated within the same educational resource with a clear reference to the image.

A clear reference to the source and originator of the image must be stated.

Theo Gray's Wooden Periodic Table

An excellent website that has all sorts of teaching applications. The link leads to a page with 'thumbnail' images of all the photographs on the site: **http://theodoregray.com/PeriodicTable/indexA.html# tabletop** (June 2004)

The Element Collection

The website for a commercial collection of the elements. Click on any of the element symbols to go to a photo of that element: **http://www.element-collection.com/html/explore_frameset2.htm** (June 2004)

The DHD Photo Gallery

A large collection of images. Note the copyright requirements carefully: **http://gallery.hd.org/** (June 2004)

Video Clips

Inserting a video clip

Video clips may be inserted into PowerPoint® slides (or Excel® spreadsheets). Formats for clips include avi, mov and mpg. Various software applications are designed to play these include Windows Media Player®, Quicktime® and Real One® and each of these formats can have a range of additional formats. Unfortunately, these permutations mean that video clips are often incompatible with the software installed on a particular computer and in addition Microsoft Office® applications aren't always set up to play all of the clip formats. This can lead to frustration.

Find a source of video clips *eg* the samples available at the RSC project website **http://www.ChemIT.co.uk**, also on the CDROM accompanying these guidelines. CDROMs such as the RSC's *Alchemy? Chemistry and Industrial Processes for Schools and Colleges* are another potential source. First establish if the clip will run on the machine by double clicking its icon. There are several reasons why a clip may not play, including the relevant 'media player' not being installed on the computer. Once satisfied that a clip will run, try inserting it into a PowerPoint® slide. The advantage of this is that the video can then be run seamlessly within a PowerPoint® show without the need to launch another application:

1. Open a PowerPoint® slide.

2. Select the Insert menu, then Movies and Sounds then Mov*ie* from file.

3. Navigate to the folder where the clip is saved, select it and click OK.
   Select whether the clip should play automatically as it appears in the slide presentation or only when activated by pointing at it with the mouse and left-clicking. Save the file

If a video clip will run on the machine, but not within a PowerPoint® slide, then it should be possible to create a hyperlink on the slide to launch the video from.

Creating a 'hyperlink' to a video clip from PowerPoint®

Any object on a slide can be made a hyperlink, and the link can be set to work either from a mouse click or simply hovering the mouse over that area of the screen.

1. With a slide in the edit view, on the Drawing toolbar select AutoShapes, Action Buttons then click the blank choice (Action Button: Custom). Left mouse click and drag a button shape on the slide. An Action Settings menu should appear, if not right click the button to display a menu and choose Hyperlink.

2. Click the Hyperlink to option and then select Other File, which is probably the last in the menu. Navigate to the video clip and select this for the link.
   OK the choice.

3. Save your work, play the slide and click the hyperlink button when it appears. This should cause a media player to open and play the video clip.

Flash® animations

Flash® animations (and less often Shockwave® animations) are very widely used within web-page-based educational materials as a means of providing animation and interactivity. Both are produced using Macromedia® authoring software. For chemistry, these animations have great potential for helping students to visualise models and reinforce concepts of what is happening at the level of particles, particularly when movement is involved eg for changes of state and interactions between the particles.

Free 'plug-in' software is required for these animations to work within a webpage, so for students to view these animations across a school or college computer network it may be necessary to liaise with technical support to ensure that the necessary software is installed. Because of this uncertainty it is particularly important to check these materials on any machines students will be using in advance.

The free Flash® player needed to view these files within a web page is available from **http://www.macromedia.com/** (June 2004)

The animation features currently available within PowerPoint® do not allow simulations of the same clarity and quality to be created. However, for simple sequential animations the simplicity and opportunity for editing offered by PowerPoint® makes animations created using it far more accessible to the majority of teachers than Flash® files are.

However, being able to incorporate Flash® animations produced by others within their lessons (assuming copyright permission has been granted) is a very useful skill for teachers of chemistry to develop.

Flash® files may be recognised by 'swf' at the end of their file name. As long as copyright permission is available, these files may be used out of their original context and integrated into a teacher's personal lesson plan in a number of ways.

Be prepared for disappointments or frustrations when trying to link to or integrate these animations. Taking materials from other resources can prove to be a hit and miss affair.

Internet sources of science Flash® animations

There are a few examples of Flash® animations which may be downloaded from **http://www.ChemIT.co.uk** and are available on the CDROM.

There are many websites that use Flash® animations to illustrate chemical processes and several of these permit their animations to be used for educational purposes, with certain conditions outlined on the websites.
Two examples (June 2004) are:
http://www.brainpop.com/

and Iowa State University Flash® Animations (copying these words into Google gives the link and is quicker and more reliable than typing the web address out)
http://www.chem.iastate.edu/group/Greenbowe/sections/projectfolder/animationsindex.htm

Link to web pages containing Flash® animations

A straightforward way of integrating a Flash® animation with a PowerPoint® slide is to provide a hyperlink on the slide that launchs an application (usually a web page) containing the animation. It is then possible to 'toggle' back and forward between the two open applications.

1. Save the animation file (eg swf ending) in an appropriate folder. Often this is the same folder as the PowerPoint® file as this makes it easier to remember to copy them both if the activity is transferred elsewhere.

2. Double click the animation file to make sure that it will run. Remember that the Macromedia® plug-in required to play the animation must be installed. Also, it may be necessary to 'associate' the file with a web page browser:

   - select the file (single left click mouse);

   - open its menu (right click mouse),

   - if Open with does not appear in the list then hold the shift key down and at the same time right click mouse;

- select Open with and choose Internet Explorer (iexplore) and also check the Always use this programme to open this type of file box;

- try to run the file again

  If the file still doesn't run and the animation is an especially useful one, then seek technical assistance. Otherwise it is more time effective to search for an alternative.

3. In PowerPoint®, select Drawing toolbar / AutoShapes / Action Buttons then click the blank choice (Action Button: Custom). Left click drag a button shape on the slide and an 'Action Settings' menu should appear, if not right click the button to display a menu and choose hyperlink.

4. Click the Hyperlink to ... choice and then select 'Other File'. Now navigate to the animation file saved as per steps 1 and 2 above and select this for the hyperlink. OK your choice.

5. Save your work, play the slide and click the hyperlink button when it appears. This should cause a web page (or even the Macromedia® player itself) to open and allow the animation to be viewed in a window outside of the PowerPoint® slide. If required, it should be possible to toggle between the PowerPoint® 'show' and the animation by holding the Alt key down and pressing the Tab key. Closing the animation window should also return the screen to the PowerPoint® show.

Embed Flash® animations

A Flash® animation may be 'embedded' within a PowerPoint® file using features already present with most versions. This approach has the great advantage that the animation is likely to run when transferred to another computer without the need to install additional software. Embedding the animation also makes it appear seamlessly within the PowerPoint® show. It also means custom annotations may be added alongside the animation.

However, the steps required to embed an animation are rather complicated and can be problematic. They are included here for those who wish to explore this method, since if successful (and it doesn't work for all Flash® animations) it is a convenient approach, particularly for files that are to be distributed.

This method requires machines that have the Shockwave Flash® ActiveX Object installed, although this is highly likely if Internet Explorer successfully loads websites that use Flash®.

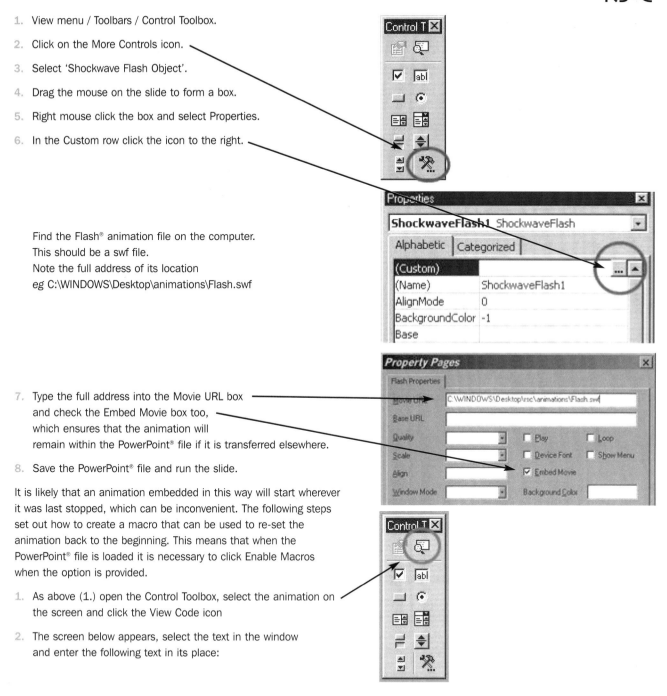

1. View menu / Toolbars / Control Toolbox.

2. Click on the More Controls icon.

3. Select 'Shockwave Flash Object'.

4. Drag the mouse on the slide to form a box.

5. Right mouse click the box and select Properties.

6. In the Custom row click the icon to the right.

Find the Flash® animation file on the computer.
This should be a swf file.
Note the full address of its location
eg C:\WINDOWS\Desktop\animations\Flash.swf

7. Type the full address into the Movie URL box
   and check the Embed Movie box too,
   which ensures that the animation will
   remain within the PowerPoint® file if it is transferred elsewhere.

8. Save the PowerPoint® file and run the slide.

It is likely that an animation embedded in this way will start wherever
it was last stopped, which can be inconvenient. The following steps
set out how to create a macro that can be used to re-set the
animation back to the beginning. This means that when the
PowerPoint® file is loaded it is necessary to click Enable Macros
when the option is provided.

1. As above (1.) open the Control Toolbox, select the animation on
   the screen and click the View Code icon

2. The screen below appears, select the text in the window
   and enter the following text in its place:

```
Sub RewindMovie()
Slide1.ShockwaveFlash1.GotoFrame 1
Slide1.ShockwaveFlash1.Play
End Sub
```

3. Select General and Declarations from the drop-down choices
   and Save

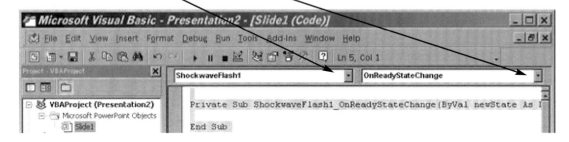

Finally a means of activating the macro is needed.

1. Draw a shape on the same slide *eg* a rectangle.

2. Right click this with the mouse, select Action Settings and in either of the Mouse Click or the Mouse Over tab, click the Run macro button and select the ReWindMovie macro for that slide from the drop-down menu.

3. Including this macro means that when the PowerPoint® file is selected an Enable Macros option will appear before the software will open the file.

Insert Flash® animations using 'add-in' software

A free software 'add-in' may be installed with PowerPoint® that allows Flash® animations to be inserted and played within a slide far more easily than the method described above. However, despite this being very straightforward in comparison, this approach has the great disadvantage that the add-in software must also be installed on the machine used to view the inserted animation in order to play it within PowerPoint®.

The free Swiff Point Player Software enabling Flash® animations to be inserted and run within PowerPoint® may be downloaded from **http://www.globfx.com** (June 2004).

Drawing chemical structures and diagrams using Microsoft applications

This section contains suggestions for using the drawing tools available within all Microsoft applications. Elaborate structures and diagrams can be built up rapidly using options in the Drawing toolbar. This is an alternative to using specialist chemistry drawing software such as ACD / Labs ChemSketch, described elsewhere within these guidelines .

The steps below indicate how a representation of methylbenzene may be developed:

1. Duplicate a new slide and type 'Drawing methylbenzene' in the title box. Click-drag the mouse over all other objects on the page to select them and then press the delete key.

2. Select the Drawing toolbar / Autoshapes button / Basic Shapes and select the Hexagon option. Hold the Shift key down to draw a regular hexagon by left click-dragging the mouse.

3. Rotate this shape (select it, Drawing toolbar / Draw button / Rotate or Flip).

4. Draw a circle by choosing the Draw toolbar / Oval button. Hold the Shift key down and draw a 'regular oval' that fits inside the hexagon.

5. Select both shapes and set line size and colour formats.

6. Select both shapes at once and align them: Draw button menu, Align or Distribute and use the Align Centre and Align Middle options as appropriate.

7. Click the Drawing line button, hold the shift key and draw a bond from the hexagon. It's the wrong thickness and possible the wrong colour!
    Select the hexagon or ring, then click the Format painter button and click the bond.

8. Click the Drawing Text box button and click-drag a box. Type 'CH3' in here (NB Ctrl + = gives subscript) and format text to suit. Ensure the fill and line colours for the box are set to 'No Fill'. Move the CH3 box next to its bond.

Grouping objects

This step is very important as it creates a single completed image, rather than a complicated assemblage of different drawing objects.

Left click-drag the mouse across all four objects to select them and then Group them as one shape: Draw button menu / Group. Position this group on the page. If drawing this in PowerPoint®, the image may be animated to appear within a sequence of events:

Although producing a structural representation such as this can be a relatively involved procedure, it only has to be followed once. When a shape has been created it can be saved, copied, duplicated (control + D), un-grouped and amended to make the structure of another aromatic molecule:

Examples, all derived from the methybenzene structure created above:

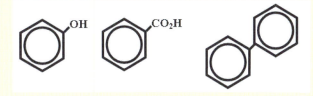

Curly arrows

These can be drawn for organic mechanisms by selecting AutoShapes then Lines then the bottom left choice. Click the mouse where the arrow is to start, let go and move to where the curve is to bend and click again, let go and move to where the arrow head is needed and double click the mouse to complete the curved line. With the line selected, click the Line

Style button, choose More Lines and set the End Style to the desired arrow.

Try combining the drawing and animation skills developed so far to create a organic mechanism sequence on the slide.

Using specialist chemical drawing software

Chemical structures can be drawn using specialist chemistry software. Once familiar with one of these packages this is a probably a quicker and more consistent way to create images of chemical structures than using a generic drawing package. However, the specialist software needs to be available on every computer used and some investment of time is needed to become familiar with techniques and skills needed.

Detailed guidelines for using one of these packages, ACD ChemSketch® , are set out elsewhere within these notes.

ACD/ChemSketch® make a free version of their chemical drawing software available from www.acdlabs.com (June 2004) and on the CDROM.

http://www.btinternet.com/~chemistry.diagrams/ (June 2004) is a collection of diagrams free to download from a website developed for and run by chemistry teachers.

Fonts

Some symbols are not always available within currently used fonts. A special RSC font is included on the CDROM and website which contains some useful symbols. See page 109.

Electronic worksheets using PowerPoint®

The PowerPoint® file e-worksheet examples.ppt (resource 33) illustrates some possibilities. These deliberately make use of only the most basic features offered by PowerPoint®. Similar activities may be created from scratch by applying the guidelines for creating a sequence of animated slides given above.

Customise a PowerPoint® template

Editing an existing presentation allows the creation of some basic review activities for learners and requires little more than text entry and pasting new images. Customising an existing format may help in gaining the confidence to create new ones from scratch. The following sections refer to the PowerPoint® file e-worksheet templates.ppt (resource 34) which is available from the project website **http://www.ChemIT.co.uk** and on the CDROM.

These guidelines assume some familiarity with the introductory sections, which it may be necessary to refer to from time to time.

Before customising the template create a copy of the file.

There are four types of review template in this presentation file, with objects appearing in a pre-determined sequence. Brief instructions for modifying each of these are given on the following pages. The purpose of each text box should be apparent from its text. Learners are advised to approach reviews like these by continually asking themselves 'what happens next?' before revealing feedback, which may be an answer or a prompt to think further.

To run any one of the four reviews and see how it works, select its name in the left screen then click the tiny icon of a screen (hovering the mouse over it displays the label 'Slide Show') below this left screen. Press the Esc key to get back to the PowerPoint® Normal View in order to customise the sides.

Think how many questions of each review type will be required and in what order they should appear. Now duplicate the number of each type required.

Slides may be duplicated by selecting them in the left screen (*ie* the whole of the slide title will be black), holding the

Control key and then pressing the D key at the same time (Ctrl+D). Change the order by selecting an individual slide, cutting it (Ctrl+X keys), selecting where you want it to appear in the list and pasting (Ctrl+V).

Delete unwanted slides by selecting them in the Outline window on the left of the Normal view screen and pressing the Delete key.

1   'Answer appears next' review

Edit the existing text, adjusting the size, style and colour of the font or fill as desired. The size of the text boxes can be adjusted by dragging their edges.

Delete unwanted text boxes by selecting them and deleting.

Add new text boxes to the sequence by selecting the previous object in the sequence and duplicating it (Ctrl+D). An exact copy should appear with the same animation but next in the sequence.

If required, insert pictures etc from elsewhere (eg the internet, if copyright permits), and then add animations and place them within a sequence of events. Refer to the relevant guidelines elsewhere if necessary.

2   'Prompt before answer' review

As above, edit the template text boxes. Do what the text in the template prompt boxes suggest ie move each of the green prompt and the yellow answer boxes over the previous text box to create an interactive sequence.

A third prompt may be created by duplicating the second one. The duplicated box should appear at the correct point in the order but it will appear on top of the answer box and so obscure the answer. To correct this:

* make sure that the Drawing menu is displayed at the bottom of the screen. If it isn't select View menu, Toolbars and Drawing.

* select the answer box, ie the one to appear on top of the prompt boxes, so that it hides them. Using the Draw toolbar, select Draw, Order, then Bring to the front.

If new text boxes have been added then check the review slide is working by clicking on the tiny Slide Show icon below the list of slide titles on the left of the Normal View school and run through the sequence.

Again, to insert pictures etc, add Custom Animations and place these within the sequence using the Order & Timing window.

3   'Label an image (ovals)' review

To customise this template the sample diagram needs to be deleted and replaced with any transparent diagram to which labels may be added for students to identify.

1. Select the image and delete it.

2. Copy the image to place on the slide and paste it on to the slide.

3. Alter the size and position of the drawing to suit needs.

Re-position the highlight colours so that they lie behind an appropriate portion of the diagram. The diagram needs to have a transparent background for this to work. If it doesn't, use the ring highlight template on the next slide.

Change the size of the highlight shape if necessary.

If some objects obscure others then choose the Draw menu, Order, Send to back or Bring forward as necessary.

4   'Label an image (rings)' review

If the image is a 'solid' picture (eg a photograph) then this template can be used to place coloured rings on top of the image.

Delete the current image on the slide, paste a new one and set its size and position on the slide.

The new image will need to be 'sent to the back' of the other objects on the slide which can be done by selecting the image, choosing the Draw menu and selecting Order, Send to back.

Position each ring and its label appropriately and type some relevant label text, re-sizing the ring and its label text box as required.

The following section illustrates how a simple labelled diagram (or sequence of textual information) might be made far more interactive through the use of navigation hyperlinks.

Using hyperlinks: active label diagrams in PowerPoint®

These are diagrams where the user can follow their own path through a PowerPoint® resource, as opposed to a pre-ordained route. There are various approaches to this, but the following points are probably relevant for most:

• Use the 'Master Slide' for text and images that will be appearing on most screens. This limits the need for duplication and also makes the resource much easier to update.

• Select Slide Show menu, Slide Transition, Advance and uncheck the On mouse click option. This means that inadvertent mouse clicks in the wrong area of the screen will have no effect.

The instructions below are for converting a diagram taken from a website of diagrams for chemistry teachers into an interactive version. Many of these diagrams already have labels and these instrctions describe how these may be replaced with new ones. Familiarity with the basics of using PowerPoint® covered elsewhere in these guidelines is assumed (eg formats for text boxes and other drawing objects, duplicating objects and slides etc).

1. Create a new PowerPoint® file, saved within an appropriate folder. Within the Outline window, select the icon for the first blank slide and duplicate this for as many times as there are different labels in the diagram plus one extra. NB use control+D to duplicate.

2. To prevent unintended jumps between slides, select Slide Show menu, Slide Transition, Advance and uncheck the On mouse click option.

3. Open the Slide Master: select View menu, Master, Slide Master. There are various default objects on the page that may be of use when developing future resources. However, for the first attempt, select and delete these so that the Master slide content is simplified.

4. Without closing PowerPoint®, copy the diagram to be used. There is a convenient internet source of these. Open a web browser and navigate to: **http://www.btinternet.com/~chemistry.diagrams/** (June 2004). To copy a diagram point the mouse at it, right click and select Copy. Return to the Master Slide, right mouse click and Paste.

NB Although this is a relatively quick way of inserting an image it is usually better practice to save the diagram image separately (using Save Picture As), eg within the folder where the resource is being created. See the guidelines elsewhere relating to inserting images within PowerPoint® presentations for details. If the inserted image is a jpg file this method (rather than just using copy and paste) will make the final file size significantly smaller.

Remember that every object on the Master slide will appear on every other slide.

5. Create a text box to give the diagram a title. Create a 'Clear Screen' navigation text box which will take the user back to their initial screen. Hyperlink this to the first slide:

   • select the edge of the text box;

   • click the Insert Hyperlink button (alternatively use the Insert menu);

   • click the Bookmark button, select Slide 1 and OK.

6. Still working with the Master slide, draw and format arrows and text boxes with appropriate indicators (eg letters or numbers) superimposing any existing labels on the diagram. NB filling the text boxes with colour can help to obscure the original label with a number or letter.

Now for the part that requires logical thinking and a sequence of repeated actions to be followed accurately...

7. Following the same procedure used with the 'Clear Screen' text box in step 5, make the first label text box a hyperlink to the second slide:

   • select the edge of the diagram label text box;

   • click the Insert Hyperlink button (alternatively use the Insert menu);

- click the Bookmark button, select Slide 2 and OK.

- save the work so far. Close the Master slide, eg by selecting View menu then Normal.

8. Select the second slide, draw a text box over the label that hyperlinks to this slide and type in the appropriate text.

9. Test that these first links work. The following should now happen if the slideshow is viewed (eg press the F5 key): the screen of labels should appear; clicking the hyperlink created in step 7 should lead to its text appearing, clicking 'Clear Screen' should link back to the opening screen.

10. If both steps work repeat step 7 to create the appropriate links to the remaining slides.

11. An additional feature to try is to include a slide that reveals all of the text labels at once:

- duplicate an additional slide at the end of the sequence; in turn, copy all the separate slide labels and paste them to this one;

- add an extra 'Reveal All' text box to the Master Slide

- and hyperlink this to the slide with all the text labels displayed.

The above is only intended to allow essential skills to be developed. In practice the design adopted will vary according to the content, format and size of the objects being hyperlinked and how the resource is intended to be used.

As with any classroom resource, teachers develop their own preferred approaches through experience of what works for them and their own students.

Using hyperlinks: providing feedback in PowerPoint®

Using the same techniques described above, hyperlinks may be used to create a resource that provides response to questions, but in a 'non-linear' way. For instance, a multiple choice activity could be created that allows a student to make any of the choices in any order through links to feedback slides from which it is possible to return to the original question slide:

Slide 1

Question text and images

Possible Answers:

A hyperlink to slide 2

B hyperlink to slide 3

C hyperlink to slide 4

D hyperlink to slide 5

If the correct choice was C (slide 4 here) then Slides 2, 3 and 5 might have the following format:

Slides 2, 3 and 5

Question text and images

Incorrect answer, for these reasons...

hyperlink back to slide 1

The correct choice could link to a slide which confirms this and includes a link to the next question:

Slide 4

Question text and images

Correct , for these reasons...

Next question hyperlink to eg slide 6
or to another PowerPoint® file

Note that when creating this type of resource it is helpful to turn off the default 'slide transition' Advance on mouse click. This is because an inadvertent click could move the viewer to the next slide, which may cause confusion. This can be done

RS•C

by selecting Slide Show menu then Slide Transition, uncheck the Advance on mouse click tick box and finally click the Apply to all button.

Creating PowerPoint® files ('edit' view)

Duplicate an object, including all of its formats and animations	Select it, then press the control key and and hold press D (Ctrl + D).

Note that if the duplicate is then positioned as required, pressing Ctrl + D again immediately afterwards will create another duplicate in the same relative position to the first. This is a very useful time saver to build up regular repeated arrays *eg* a lattice of particles.

Copy a selected object	Ctrl + C keys
Paste a selected object	Ctrl + R keys
Cut a selected object	Ctrl + X keys
Move a selected object	Arrow keys
Nudge a selected object	Ctrl + Arrow keys
Select several object at once	Hold Shift key, click each in turn
Run a slideshow from its first slide	F5 key

Typing reaction equations:

Subscript on / off	Ctrl + =
Superscript on / off	Ctrl + Shift + =
Reaction arrow (doesn't always work)	Type: − −> to get: →

During a PowerPoint® 'show':

Reveal the keyboard shortcut menu	F1 key
Switch to the edit view from a slide show	Esc key
Advance a slideshow	Left click the mouse; press the space bar; press any of the return (enter), right arrow or down arrow keys.

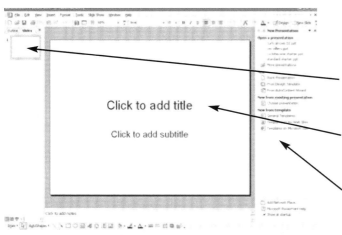

Reverse a slideshow or 	Press any of the backspace, left arrow up arrow keys.
Go to a particular slide (its number needs to be known in advance)	Type the number of the slide then press the return (enter) key.
Turn the screen black or white (eg to hide it or to use an OHP)	B or W key Press B or W again to restore the screen.
Turn the pointer to a pen Turn the pen to a pointer	Control + P Control + A
Erase pen annotations	E key

Questions and Answers

A useful website devoted to offering advice about all aspects of using PowerPoint® is found at:
http://www.rdpslides.com/pptfaq/ (June 2004)

Using PowerPoint 2002® (XP)

The following sections introduce PowerPoint XP®. This follows the same sequence as the introductory guidelines for the earlier versions of PowerPoint® described above.

Essentially the same features are available within both versions, but some are accessed and displayed differently. It is expected that once a reader wishing to use XP is familiar with these introductory pages they will be able to apply the guidelines written for the 2000 version of PowerPoint® to XP. The final part of this XP guide describes some useful features that are not present in the 2000 version.

Creating a new PowerPoint XP® slide

If PowerPoint 2000® or earlier is being used turn to the relevant section. Note that for most of these instructions there are alternative ways of achieving the same result. For instance, many actions that use the toolbar menus can also be carried out through the mouse right-click menu. Colleagues (and students) are good sources of alternatives, especially the 'shortcuts' that save time. Individuals will find the approaches most appropriate to their needs through exploring the alternatives.

Opening a new PowerPoint® file and screen layouts

Launch PowerPoint®. If a 'Blank Presentation' isn't on the screen open a new PowerPoint® 'slide show' by selecting File menu, New. The screen shown left should appear.

This is Normal view. In particular note the three labeled parts of the screen. If the Task Pane isn't visible select it from the View menu.

On the left
Provides a summary of the whole presentation, either as Slides (pictures) or Outline (text). Click the Outline tab.

Central
The current slide from the presentation which can be edited in this view.

On the right
The Task Pane, from where various formats can be added to the objects on the slide.

RS•C

The Task Pane

The Task Pane is the major difference in screen layout from previous versions, but note that most features contained within this are available through the toolbars and menus described in the notes for the earlier versions of PowerPoint® elsewhere in these guidelines.

There are various windows that can be displayed as the Task Pane. One of the most useful is Custom Animation. Open this by choosing it from the drop-down menu next to the title at the top of the Task Pane.

The following steps are a guide to producing a title, heading, reaction reactants and products which appear on the screen in sequence controlled by the mouse or keyboard.

> Note that should you make a mistake at any point, clicking the Undo button (backward curly arrow) will restore the previous setting.

A PowerPoint® file named Sample PowerPoint.ppt (resource 48) contains slides created following these introductory guidelines for comparison. This is available to download from the **http://www.ChemIT.co.uk** website and on the CDROM.

Note that these instructions are intended to support the development of confidence in working with PowerPoint®. There are many different ways to achieve the same results and it is not suggested that the methods here are any better than the alternatives. Individuals will find the ways that best suit through practice, exploration and picking up tips from others.

Select the 'File menu' (top left) and 'Save As' to reveal the window shown on the right.

Click the arrow to the right of the 'Save in' box to reveal a list of locations.

Navigate to the 'My Work' folder (or to an alternative specified for the computer you are using).

Type a sensible filename. ——————

Click Save ——————

Adding text boxes and animation

1. Either click within the blank 'Click to add title' text box or immediately to the left of the slide picture in the Outline view – both have the same effect.

   Type 'Chemistry Presentation' and a title text box will automatically appear on the slide.

   Click the edge of the title text box once and its border will be highlighted.
   Try clicking once inside the box and compare the difference.

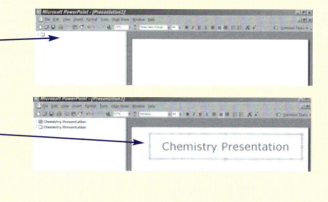

2. With the edge selected (but not any of the re-sizing circles) try dragging the text box around the screen. Then hover the mouse over one of the re-sizing circles and compare the effect of dragging these. Note that the corners maintain the proportions of the box whereas the middle of the sides distort the shape.

3. Add some relevant text (eg 'Industrial preparation of hydrogen') to the subtitle text box (ie 'Click to add subtitle). If there isn't a subtitle box on the screen create a new text box using the button on the Drawing menu below. Remember, the Undo button reverses any unwanted changes.

   This title appears as a subtitle in the Outline window. If this level of information isn't required right click the Outline window and select Collapse.

   Note that 'hovering' the mouse over a button for a few seconds usually displays information about what the button does.

   With the edge of the text box selected, change the font and font size of both boxes by making choices from the menu at the top of the screen

   Open the Drawing toolbar by selecting View Menu, Toolbars, Drawing. This will appear along the bottom of the screen.

   With the title text box selected (remember click its edge), use the Drawing toolbar button to change the Font Colour.

   To view the current slide sequence, use the Slide Show button:

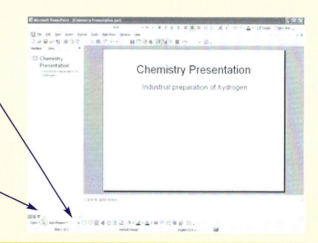

4. Use the appropriate button from the Formatting menu at the top of the screen to Left Align the subtitle text.

5. Save your work (File menu, Save).

6. Click the edge of the subtitle text box to select it and then click the Slide Show menu and select Custom Animation to open the Custom Animation Task Pane, on the right of the screen:

Click the Add Effect menu,
choose Entrance
and select an animation from the list
eg Appear.

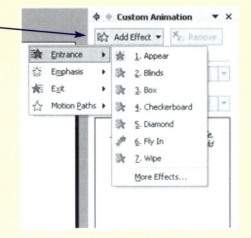

Note that the Custom Animation window allows the properties of an individual animation to be set and also allows the properties of several to be linked eg two animations for separate objects can be made to happen simultaneously. These effects are described in more detail at the end of this XP section.

7. Save the work again: click the Save button in the Standard toolbar.

Test the simple sequence created so far. There are various ways of doing this, for instance pressing the F5 key will run any show from the its first slide.

It is often useful to view the sequence for the slide being worked on. One convenient means of doing so is to click the small Slide Show button towards the bottom left of the screen (see previous page). It looks like this:

Return to the edit view from the slide show by pressing the Escape key.

Now create text boxes for the reactants and products of the reaction.

1. Click the edge of the subtitle text box in order to select it, hold the Ctrl key down and then, without releasing Ctrl, press the D key. This should duplicate the title text box, including its animation effect.

> Note: Ctrl+D can be used to duplicate a variety of PowerPoint® objects, including whole slides selected in the Outline window, text boxes and other drawing objects. This is an extremely useful time saver as it means that once an object has been formatted a duplicate with the same features can be made instantly.

2. Use the mouse to select all the text within the duplicated text box and type the formula CH_4. To type the subscript, before typing the '4' hold the Ctrl key down and then, without releasing Ctrl, press the = key. This sets the following characters as subscript. Repeat this (*ie* Ctrl with =) to turn subscript off.
   NB To create a superscript hold the Ctrl key down and then, without releasing Ctrl, press both the Shift and = keys. To turn superscript off repeat this (*ie* Ctrl with Shift and =).

3. Finish off the reactants by typing '+ H_2O (g)'. To create a reaction arrow try typing – –> (dash dash greater than) which may automatically become →.
   Position the text box on the slide by clicking the edge and using 'drag and drop'. If this doesn't work, use the arrow button on the Drawing toolbar, select the arrow and give it a custom animation (point 4 above).

4. Save the work.

5. View your presentation so far by pressing the F5 key. There are several ways to run the animation steps:

   • left click the mouse; press the space bar;

   • press any of the return (enter) or right arrow or down arrow key.

> Note that pressing any of the backspace, left arrow or up arrow keys reverses an animation, a particularly useful teaching feature.

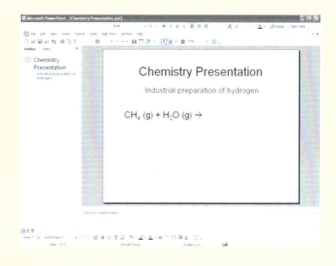

RS•C

6. Duplicate the reactants' text box and replace the reactants with the products:

   Move the product box to an appropriate position on the slide.

   Save the work and then view the animated presentation by pressing the F5 key or the Slide Show button:

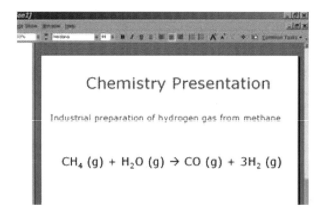

Chemistry Presentation

Industrial preparation of hydrogen gas from methane

$$CH_4 (g) + H_2O (g) \rightarrow CO (g) + 3H_2 (g)$$

Some useful features of PowerPoint 2002® (XP)

Everything described above is possible with previous versions of PowerPoint®, although the appearance of some features (eg the Custom Animation window) is different in XP. However, there are several animation options that are not available in earlier versions. Some of these are very effective when creating materials to support science teaching.

Some examples:

Motion Path Animations

It is possible to make one or more objects appear to move using pre-2002 versions of the software, but it is rather complicated and the results are confusing to edit. The Motion Path animation option in XP makes this very straightforward:

1. Select an object on the slide. Open the Custom Animation window in the Task Pane and click the Add Effect button.

2. Choose Motion Path and Draw Custom Path / Curve. Note that there are several pre-set path options.

3. Left click the mouse once where you want the path to start.
   Note that if an object is not to jump abruptly across the screen the path needs to be drawn from where the object is placed on the screen.
   Left click once again at each point the curve should bend. Double left click to complete the path.

4. Select the speed at which the object should move.

Several objects animated at once

It is possible to have several different objects' animations starting at the same time. For example two atoms could have motion paths set to move them towards each other and collide:

1. In the Custom Animation window, make sure that the actions that are to start simultaneously follow each other in the Order sequence. If any do not then select them and use the Re-Order arrow keys.

2. Select the second item of those to start simultaneously, click its drop down menu arrow and select Start With Previous.

3. Repeat step 2 with as many actions as required.

More than one animation for the same object

It is possible to link more than one animation to the same object.
Select the object on the slide and then click the Apply Effect button etc.

Drop-down menu form fields in Word® documents

These can be included using the Forms toolbar. An example to refer to is atomic structure.doc (resource 10).

Open a new Word® document and save this to an appropriate place using
File menu, Save As, give the file a sensible name and then click OK.

Select the View menu then Toolbars then Forms to display the Forms toolbar.

Hovering the mouse over each option will reveal a label describing the type of Form it will place on the page.

The Drop-Down Form Field option allows text to be selected from a list and placed at a specified point within a document:

Place the cursor where the drop-down menu is required and click the third button from the left (Drop-Down Form Field). A grey rectangle should appear. Point at the grey rectangle with the mouse and double click it to display the Drop-Down Form Field Options window shown to the right.

Type an item of text to include as a choice in the drop-down item box and click the Add button. The order an item appears in the list may be changed by clicking it and using the Move buttons.

If the space where the selected text is to appear is required to be blank when the document is opened, type a few spaces in the Drop-down item box, Add this and move it to the top of the list.

The font size, style etc for the text that will be placed in the document by the form can be set by using the mouse to select the form on the page and then using the Formatting toolbar.

The section on the next page describes how to 'activate' the form *ie* to make it available to students when they open the document.

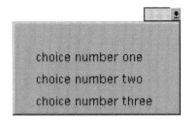

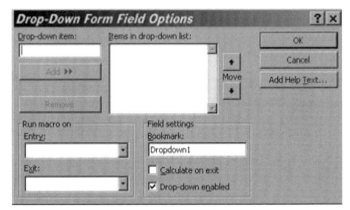

Protecting the document (activating the forms)

> When all the form fields and the rest of the document are set up, select the right-most button in the Forms toolbar *ie* the Protect Form button, which looks like a padlock.

This 'locks' the document so that the only items that can be changed are entries into the various forms on the page.

Text form fields

These can also be included using the Forms toolbar. They allow students to type text (*eg* their name in order to identify themselves if the work is to be handed in) even when the document is protected to activate the Forms. To add a text form make sure the document is unprotected (*ie* click the Protect Form button), place the cursor where the form is to appear then click the ab button to the left of the Forms toolbar. To activate the text form (and any other form in the document) click the Protect Form (padlock) button.

Check Box form fields

These allow an 'X' to be placed against an item on the page so, for instance, could be used for multiple choice style questions.

Editing existing Word® e-worksheets that use forms

Create a new document from scratch, or use an example of an e-worksheet that makes use of this feature available on the ChemIT website or CDROM as a template from which a new resource may be created. The content of an existing drop-down menu may be edited by clicking the Protect Forms button in the Forms menu and double-clicking the form to be amended.

Adding forms to word-processed files to create electronic resources

Several teachers have suggested that adding forms to existing word-processed worksheets previously used as handouts allows these materials to be converted to an electronic resource. This is a relatively quick way of adding value and flexibility to an existing resource.

Drag-and-drop objects around the page

This feature can be used for independent study e-worksheet activities but is also effective displayed on a large screen in the teaching room. It mimics activities found in several commercial packages, however creating these in Word® or Excel® has the advantage that the materials can be edited and customised. An example using Word® is distillation.doc (resource 12).

Check that the Drawing menu toolbar is displayed at the bottom of the screen. If it isn't, then display it by selecting View menu, Toolbars then Drawing. As elsewhere, what any particular button in the menu does is revealed by 'hovering' the mouse pointer over it for a few seconds.

> Note for Microsoft XP® users:
> Unlike earlier versions of MS Word®, XP may automatically place a 'drawing canvas' around any Drawing menu objects placed on the page. If this feature is not required then it can be turned off by selecting Tools menu, Options, the General tab and unchecking the option labelled 'Automatically create drawing canvas when inserting AutoShapes'.

To permit objects to be positioned precisely anywhere on the page, turn off the drawing grid: in the Drawing toolbar select Draw menu, Grid and uncheck 'Snap objects to grid'. Note that there may be times when it is desirable to have objects 'snap' to a grid, in which case the Grid option may be used to control the size of the grid.

> Any object created using the Drawing toolbar can be 'dragged and dropped' to a new position using the mouse or selecting the object and using the arrow keys.

Some possible uses

- classroom 'interactive board' exercises or independent study activities;

- assembling and labelling apparatus or diagrams;

- re-organising matching pairs *eg* definitions; diagrams and labels;

- assembling reaction equations or mechanisms.

A problem that often arises when text boxes are used in this way is that the user inadvertently selects and sometimes deletes the text content as they attempt to drag-and-drop the box. The next section sets out an approach that avoids this.

Creating a drag-and-drop picture containing text

Text can be made 'moveable' in this way by typing into a text box. However, there is a problem in that only the narrow edge of the box can be picked up and if the mouse is clicked into the box then the contents can be altered. This can be avoided by converting the text box into a picture, although this is a more complicated process and produces larger documents.

This is a multi-step process which needs working through a few times to become fully confident in its use.

It may be preferable to format an image containing text exactly as you want it before creating the object that can be dragged-and-dropped. It is more difficult to change its appearance after this.

1. Create a text box and type the required text using sub and superscripts, font, style and colour formats as required. Set the formats for the line and fill colours for the box, these may both be left clear.

   It may be desirable to create more complicated objects to drag and drop around the page by grouping the text box with an autoshape (oblong, oval etc):

2. Select all of the objects to be grouped by using the Draw toolbar, Select Objects button or by holding the shift key down and clicking each object in turn.

3. Select the Drawing toolbar, Draw menu, Group. The new moveable object will need to be formatted to sit in front of other objects on the page. Select it by pointing the mouse at it and right-clicking, then choose Format Object, Layout, *eg* In front of text.

4. When confident that an object that includes text is completed, copy (*eg* Cntrl+C keys) the text box (possibly grouped with other objects), then create the Picture version by selecting Edit menu, Paste Special, Picture.

5. If anything is wrong with the picture, delete it and edit the original copy. If the picture is appropriate delete the original. If the picture needs to be changed in the future it is usually quicker to create it from scratch rather than editing it.

6. To make it possible for the picture to be dragged and dropped anywhere in the document, point at it, right click and select
Format Picture, Layout, In front of text.

Drag-and-drop objects using PowerPoint®

When in 'presentation' mode PowerPoint® does not currently allow objects to be moved around on the screen. However, several teachers have reported that in order to feature 'drag-and-drop' during a classroom lesson they will switch to a pre-prepared PowerPoint® slide in the Normal (*ie* editable) view. The advantage here is that it is more straightforward to 'escape' a slideshow than to open Word® or Excel®.

Typing chemical symbols and characters

Font Symbols and Characters

Many useful symbols and characters can be found in other fonts, eg Δ, δ, λ, ! or ° can be incorporated into a document, (if they are 'True Type' fonts) but they will not appear when the file is opened on a different computer unless that computer also has the required fonts installed or the fonts are saved with the document.

In order to be certain that the required fonts are available wherever the document is used, the font should be saved with the document.

To save the characters or fonts with a document, open the Save As dialog box in the File menu, and find either general options or save options in the Tools drop down menu.

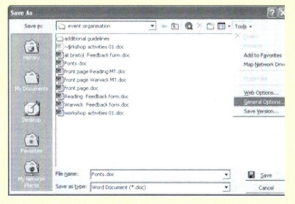

A new Save dialog box opens and a number of useful options become available. Clicking the Embed True Type fonts option saves the whole font with the document. This enables a second box called Embed characters in use only.

This option keeps the amount of information saved with a document to a minimum and is a sensible choice if only a few characters are used.

Specialist chemical fonts

There are several specialist chemistry fonts available, many of which may be downloaded for free via the internet. These are useful for specialist chemical symbols such as equilibrium and standard states eg $\rightleftharpoons$ or $\ominus$. It is worth investigating more than one of these fonts, as they offer different features.

1. Find a chemical font. A copy of the Royal Society of Chemistry font is available on the enclosed CDROM, which includes several useful symbols. There are several similar fonts available eg

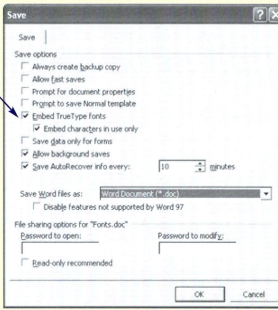

ChemFont97 which is free from **http://www.scs-intl.com/** (June 2004), subject to a copyright statement. Alternatively try searching the web for 'chemistry font'.

2. Save this in the font folder of the computer, which should be in the Windows folder. If this folder isn't there, try using the 'Find' option in the Start menu and search for 'fonts'.

3. Open a document or file, select the Insert menu then Symbol. Scroll down the menu to find the chemical font that has been installed and select the particular symbol required.

4. Remember to save the specialist font within the file or document (see previous page) so that these will be available even if this is opened on a computer that does not have the font installed.

Formulae and reaction equations

As described elsewhere, keyboard shortcuts may be used to make the typing of formulae and equations very straightforward:

turn subscript on or off	Ctrl and = keys pressed together
turn superscript on or off	Ctrl and shift and = keys pressed together
to produce → automatically	type – –>

However, this does not allow the superimposition of sub and superscripts.

Equation Editor

This feature is often included with Word®. If it has been installed it is available within the Insert menu:

Position the cursor where the equation or formula is required, select the Insert menu, Object, then Microsoft Equation. This feature may be cumbersome to use but allows superimposition *eg* mass superscript over atomic number subscript *eg*

$$^{12}_{6}C$$

The Chemistry Formatter freeware

Free software that allows specialist chemical content within Word® or Excel®. An example is the Chemistry Formatter add-in which is available from **http://spectrum.troyst.edu/~cking/ChemFormat/** (June 2004). This software installs a toolbar which automatically converts typed text into chemical formulae.

Creating hyperlinks in Word®

Text (or other objects) may be formatted so that a single mouse click will move the view to elsewhere within the same document or open a different application altogether *ie* hyperlinks may be created. NB links to places within the same document are referred to as bookmarks.

When a link is clicked a navigation toolbar will automatically appear towards the top of the screen. This includes a button which allows the path taken to be reversed. Creating hyperlinks allows:

• structure, as the designer can provide a framework that guides a user to different locations with advice about what can be found if a link is followed *eg* a 'curriculum map' for a topic;

• more complex resources, as a range of activities may be connected *eg* inter-linked Word®, Excel®. PowerPoint® and web pages;

• user flexibility, as the opportunity to navigate allows choices about the route taken through a resource;

• feedback, as a link may indicate whether a choice is correct or not, with a reason why.

Creating hyperlinks follows broadly the same pattern in all Microsoft applications, including the web page design application Front Page.

To link to a different file or live website:

1. Select the object that is to be made into a hyperlink *eg* some text, a picture or any drawing object such as a text box.

2. Open the Insert Hyperlink window, there are several ways:

   • select Insert menu, Hyperlink

   • or press both the Ctrl and K keys at once

   • or right click the mouse and select Hyperlink

   • or click the Insert Hyperlink button on the Standard toolbar.

3. To link to another file use the appropriate Browse for... option. To link to a website whilst not connected to the internet, enter the URL (address) of the web page into the relevant space. If connected to the world wide web click the Web Page button to open a web browser and navigate to the page to be linked, copy the URL from the address (using the Ctrl and C keys pressed both at once), move back to the Hyperlink window and paste the address (Ctrl and V keys).

To link to text elsewhere in the same document *ie* create a Bookmark:

1. Select the text target for the link *eg* a subheading elsewhere in the document.

2. Select Insert menu, Bookmark, give this target a name and click Add.

3. Follow steps 1. and 2. above and then use the Bookmark option to make a left-mouse-click link to any text previously bookmarked.

RS•C

Creating web pages using Word®

Web pages with links to resources are a convenient way of making resources accessible to learners. An 'intranet' of materials can be distributed via a computer network or disk.

If there is no access to specialist web authoring software (eg MS Front Page®, Macromedia Dreamweaver® etc), or no expertise in using such software, it is possible to create web pages by saving Word® documents in this format. Although effective there are technical reasons why this isn't the best way to create such a resource, particularly if it is intended to edit pages at a later date.

1. Create a sensible folder structure to store resources and web pages. For example:

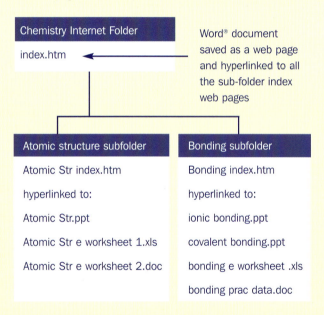

Chemistry Internet Folder

index.htm

Word® document saved as a web page and hyperlinked to all the sub-folder index web pages

Atomic structure subfolder

Atomic Str index.htm

hyperlinked to:

Atomic Str.ppt

Atomic Str e worksheet 1.xls

Atomic Str e worksheet 2.doc

Bonding subfolder

Bonding index.htm

hyperlinked to:

ionic bonding.ppt

covalent bonding.ppt

bonding e worksheet .xls

bonding prac data.doc

2. Open a new Word® document and use File menu, Save As, to select or create a suitable folder, type an appropriate name in the File name box , choose Web Page (*.htm,*html) from the Save as type menu and then OK. Note that although the file created can still be edited in Word® it is no longer a 'doc' file but now has the extension 'htm'.

3. Type into this document as you would normally and create hyperlinks (previous section) to resources.

4. Create different web pages from other Word® documents in the same way as 2 and create hyperlinks to connect these. Always include a 'back to main index' link on each page so that users can always get back to the top level.

When any of the web page files created in this way are double-clicked a web browser is launched and the file opens as a web page **which therefore cannot be edited**. To edit a web page created in Word®, launch the Word® application and use File menu, Open.

Using Excel®

Modifying Excel® template interactive worksheets

Using Excel® has the advantage that feedback can be provided as to whether a choice is correct or not. As creating these is a little more complicated than in Word® or PowerPoint®, templates are provided that require text entry only. Several templates are provided via the **http://www.ChemIT.co.uk** website or the CDROM provided. Open any of these, clicking to 'enable macros' if this is requested.

These approaches were developed during the Teacher Fellow project year (2002–2003). An alternative approach to creating this type of interactive worksheet with Excel has been added to the CDROM resources at the time of writing these guidelines. Although this means it hasn't yet been tested extensively it offers a more straightforward and flexible approach. It makes use of 'Lookup Tables' which allow far more choice with the feedback provided. It doesn't necessarily use 'macros', thus avoiding the need to 'enable' these in the particular version of Excel being used. It also includes a method for counting scores, either for the student or hidden so that the teacher can use it to mark student work.

Please read the following carefully before trying to edit a template:

> Each template contains a 'worksheet' spreadsheet which the student interacts with and an 'answers' (or 'responses') spreadsheet which controls what appears on the first sheet. The 'answers' (or 'responses') sheet contains coloured cells where new text may be typed to customise the responses. To move between these two sheets, use the tabs at the bottom of each sheet. Note that these tabs might not be visible if the display has been set to 'Full Screen'.
>
> Before starting to edit any template use Save As in the File menu to make a copy to work on and avoid overwriting the template.
>
> The next page includes stepwise instructions for editing two examples.

multiple choice template.xls	multiple choice questions and answers, resource 28
drop down template.xls	drop down menus ('combo boxes'), resource 30
type answer template.xls	only accepts the exact text required, encourages correct use of terminology, resource 32
self test quiz template.xls	uses drop down menus and click buttons to explore a topic, resource 49
lookup_template.xls	uses 'Lookup Tables' and can provide scores, resource 69

Creating a new resource using the 'drop down' template

Open the template called drop down template.xls (resource 30). If a window about macros appears click Enable macros. Create a copy of the file eg by using File menu then Save As to save a new self-test spreadsheet with an appropriate name.

Ensure the Formula bar is visible above the sheet. This indicates the contents or formula of the cell currently selected. If the Formula bar isn't visible select Tools menu, Options, View tab then click in the Formula bar tick box.

1. In the e-worksheet spreadsheet click on any of the question cells, then click within the Formula bar and edit the text. Note that images can be copied and pasted onto the spreadsheet to extend the range of questions.

2. In the answers sheet any of the highlighted cells can be amended. Enter new choices for possible answers, these can be sentences as well as single words. If Answer 1 is left blank then no choices are visible when the e-worksheet is first opened.

3. Below the possible answers, enter the number that identifies the correct one.

4. Repeat steps 1–3 to amend each question.

   If any of the feedback messages need to be changed their text can be altered.

RS•C

Creating a new resource using the 'type answer' template

Open the template called type answer template.xls (resource 32). If a window about macros appears click Enable macros. Create a copy of the file *eg* by using File menu then Save As to save a new self-test spreadsheet with an appropriate name.

Ensure the Formula bar is visible above the sheet. This indicates the contents or formula in the cell currently selected. If this isn't visible select Tools menu, Options, View tab then click in the Formula bar tick box.

For each question in the e-worksheet sheet, click on the question text, then click within the Formula bar and edit its text. Note that images can be copied and pasted onto the spreadsheet to extend the range of questions.

In the answers sheet any of the highlighted cells. Enter the correct text in the 'Answer' cell in row 6. Then copy and paste this between the quotation marks in the cell immediately below in row 8.

Note that the text must be entered precisely (although upper or lower case are not distinguished between) *eg* an additional space will lead to the answer being identified as incorrect. Users may wish to point this out to their students.

Creating a new resource using the 'Lookup Table' template

Open the template called lookup_template.xls (resource 69). Follow the instructions within the text box on this spreadsheet.

Creating interactive worksheets using Excel®

This section assumes the reader already has some familiarity with Excel®.

The following guidelines describe how the types of resource made available as the templates, and described in the previous section, might be created from scratch. Note that as well as text they may incorporate diagrams, drawings, photographs or video clips.

Appearance of the spreadsheet

The appearance of existing spreadsheets (including the templates) or new ones may be changed to suit their purpose. They do not need to be displayed in the traditional spreadsheet arrangement of cells in rows and columns.

The format of the text in cells may be set by selecting the cell and using the Formatting toolbar as required.

To set a background colour for the whole spreadsheet: first select all of the cells in the sheet by clicking the junction between the column and row header labels *ie* above 1 and to the left of A. Then select Format menu, Cells, Patterns and choose a colour.

To hide row and column headers, gridlines and other features that might be distracting to students, select Tools menu, Options, View and then un-check the appropriate boxes.

The text for questions etc. can be typed into spreadsheet cells.This may, however, restrict the layout, particularly as more questions are added. An alternative is to use a Text Box (from the Drawing toolbar) and position this as reequired on the sheet.

If the Text Box is typed in Word® or Powerpoint®, subscripts and superscripts can be added using keyboard shortcuts (Ctrl + and Ctrl + shift + = respectivey), which can be convenient. Irritatingly, copying and pasting a Text Box created like this into Excel® leads to loss of these features. One method of avoiding this is to copy the Text Box but use Edit menu, Paste Special and choose one of the Picture options. The disadvantage of this is that the pasted image cannot now be edited.

Hiding spreadsheet cells

Rows or columns of cells, or the whole sheet, may be hidden from view eg it may be desirable to hid formulae or tables that generate feedback.

1. Select the block of cells to be hidden from view (NB to select a range of columns or rows, click a column letter or row number and then drag across the other letters or numbers to be hidden to the last one and then release the mouse button).

2. Select Format menu, Column, Row or Sheet, Hide.

3. To reveal hidden cells click the column (or row) letters (or numbers) on both sides of the hidden cells then select Format menu, Column or Row, Unhide. To reveal a hidden sheet select Sheet, Unhide and select it from the list.

Protecting spreadsheets

This prevents a user changing the appearance or function of a spreadsheet and also from looking at hidden cells. It can make use of a password which prevents a user removing the protection. Obviously, if a password is forgotten or mislaid then the spreadsheet is protected from its designer too, so take care with this feature.

It is essential to leave cells which require a student to enter or change data unprotected (or 'unlocked').

1. Select any cells requiring data entry from the student.

2. Select Format menu, Cells, click the Protection tab and uncheck the Locked choice.

3. Select Tools menu, Protect Sheet (or Workbook if preferred) and click OK if required to add a password.

4. To unprotect a sheet (or workbook), select Tools menu then Unprotect...

Creating a 'Lookup Table' e-worksheet (including mark counter)

There is a template for this on the CDROM and ChemIT website, lookup_template.xls (resource 69). It may be helpful to refer to this when reading through the following guidelines for the first time. Immediately after these instructions there are screenshots of the final spreadsheet produced.

Identify an area for question. Identify an area where the student can respond to the question and where feedback will appear. Identify an area where the teacher can enter feedback possibilities and also, if required, an area to record marks and progress. For example, the spreadsheet could have the following design layout:

Report of progress		Hidden cells	Hidden cells
Questions	Student response and feed-back for this	Feedback possibilities entered by teacher	Record of progress

1. Leaving some space for monitoring progress, type the first question, either in the spreadsheet or within a text box (see the Appearance of the spreadsheet section above). For instance:

   Q1 Which of the following is a group 1 element?
   A   Chlorine
   B   Potassium
   C   Aluminium

2. Type a two column table of responses within the 'Feedback possibilities' area, where each of the possible choices occupies a different cell in the first column and each corresponding response occupies a different cell in the second column eg in the cell range K7:L10.

A	No, this is from Group 7 and is a non-metal
B	Yes, this is from Group 1 and is a reactive metal
C	No, although this is a metal it is from Group 3, not Group 1
type here	Type your response into the blue cell

NB the first column must be in ascending order ie in this example alphabetical. The last row places a comment in the sheet prompting a student response.

Further responses may be added as extra rows, as long as the first column remains in ascending order eg D and E plus feedback could be added between the 'C' and 'type here' rows.

3. Decide which cell is to be designated for students to type their response to the question eg cell H10. This might be made clear by giving it a fill colour eg by using Format menu, Cells, Patterns then selecting a colour.

In order for feedback to be associated with what a student types, enter a 'Lookup' formula in a cell adjacent to the cell where the student responds. For this example, suppose the student data entry cell is H10, the table occupies the cell range K7:L10 and the feedback statements are in the second column of the table. Typing the following formula into cell G12 will produce feedback in G12 corresponding to the response typed into H10:

=VLOOKUP(H10,K7:L10,2)

NB the 'V' is an instruction to match values in vertical columns of the table.

Saving the spreadsheet with 'type here' typed into H10 will set it so that when opened there is an instruction prompting the students to type into the appropriate cell. If required this instruction can be customised for each question by editing that question's own Lookup table.

In the example, typing 'A' should lead to the text 'No, this is from Group 7 and is a non-metal' being displayed in G12. Upper and lower case are treated the same ie 'A' gives the same result as 'a'.

The formula will match closest fits eg table 'bat' should lead to the B response being displayed, typing anything beginning with a letter later than C in the alphabet will lead to the C response being displayed. Consequently it is worthwhile giving careful thought to the instruction for what should be typed by the student.

Monitoring progress: adding a mark counter

In a cell to the right of the table for the first question enter the number 1, eg enter 1 in cell N7 then enter a formula in cell F3 to add up all the entries for all the questions, which in this case will be in the N column: =SUM(N5:N500). Consequently the number in F3 will be the total number of questions. Type some appropriate text next to F3 eg 'Total number of questions'.

In cell O7 type a formula that displays '1' if the correct feedback is displayed in cell G12 and '0' for any other response. In the example cell L8 of the Lookup table contains the feedback to the required response so a suitable cell formula would be: =IF(L8=G12,1,0).

Now enter a formula in cell F2 that displays the total in column O, which will be the number of questions answered correctly: =SUM(O5:O500) and type some appropriate text next to this eg 'Correct responses'.

When the first question is duplicated (see below) all of the formats needed to monitor progress through the questions will also be copied, but note that it is essential to update the cell reference to the correct response in the new question's own Lookup Table.

Finally, set the spreadsheet to allow the progress information to always be on the screen as the student scrolls down through the questions. Click the row number below the lowest row that is required to remain in view, eg click on the number for row 5, which selects all of row 5, and select Window menu, Freeze Panes. This keeps the first five rows in view, ie the progress monitoring information, whilst scrolling down the rest of the sheet.

Creating further questions

When the formulae have been tested and the design meets requirements it is straightforward to create further questions (and to monitor progress through them) by duplicating the first one.

Select all of the rows containing data relevant to the first question (by clicking the number of the first row required and dragging down to the last), copy these rows, click in the column A cell below Q1 and paste. Note that if the question text was created using a text box saved as a picture, then a copy of this cannot be edited. In this case a new text box needs to be created and placed in position.

It is straightforward to change the response text in the second column of the table and this allows specific feedback to be made available related to each response the student chooses. Not only can the student be told a response is correct or incorrect, they can be also be given information about why or advice for further action. Note that the formula in column O (*ie* indicating whether the correct response has been chosen) needs to be updated to point to the cell in the Lookup table with the correct response.

Hiding the data entry cells and protecting the worksheet

Select the columns containing the tables that the Lookup formula refers to and use Format menu, Hide. If the sheet is to be protected, *eg* to prevent anything being altered, inadvertently or otherwise, ensure that the cell the student types in is 'unlocked' by selecting the cell then using Format menu, Cells, Protection and uncheck Locked. Then use Tools menu, Protect Sheet (or Workbook). If the progress monitoring is not to be accessible by the student, *eg* if a teacher wants a student to work through without knowing which answers are correct and has designed the feedback accordingly, then the first few rows may also be hidden. In the example described here rows 1 to 4 might be hidden.

Screenshots for the example

These views illustrate what the e-worksheet could look like following the previous instructions. They are for the stage after the first question has been duplicated once and the correct response 'counter' reset to account for a different choice made in question 2. The second view displays the formula used.

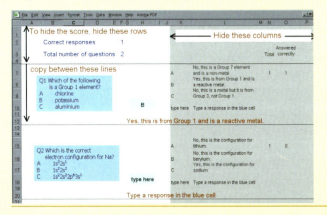

Creating a 'drop-down menu' e-worksheet

These use what are referred to in Excel® as Combo Box Forms.

To examine how an existing combo box is set up, *eg* in the drop down template.xls (resource 30), open the spreadsheet, point the cursor at a form, right click with the mouse and select Format Control from the menu that appears.

The following steps provide guidance for creating drop-down menu worksheets from scratch. This is significantly more involved than simply re-typing question and answer text into the template.

Use two Excel® sheets within a workbook, designating one the 'User' sheet and the other an 'Admin' sheet.

Decide on the question and type this into a cell as Q1 on the User sheet.

The next step is to create the cells that will provide the information for the 'combo box'. Sensibly these cells will be on the Admin Sheet *ie* not on the User Sheet, where Q1 appears. For example, to create a list of particles type the following on the Admin Sheet:

RS•C

	Q1
1	
2	atom
3	ion
4	molecule
5	electron

The numbers indicate the drop-down list order.

Note that in this example the first choice for Q1 is left blank. This is so the drop-down box is empty when the student opens the worksheet.

Elsewhere on the Admin sheet, type the following labels into separate cells *eg*

	B	C
10	correct answer	
11	user answer	
12	user difference	=(correct cell ref – user cell ref) *eg* =C10-C11

Now create the Combo Box Form next to the Q1 cell on the User Sheet.

1. If the Forms toolbar is not already displayed then select View menu, Toolbars then Forms. Remember 'hovering' the mouse cursor over any toolbar button reveals its purpose.

2. Click the Combo Box button. Click on the spreadsheet where this is to be placed and drag the shape to the required size *eg*

3. Point at this with the mouse and right click to view a menu.
   Select Format Control, Control and click into the Input range: box. Click the tab for the Admin Sheet. Find the

cells that are to appear in the drop-down list (*ie* 'blank' to 'electron' if the example is being followed), click the first, and whilst holding the mouse button, drag down to the last in the list *eg* $B$3:$B$7 in the box means the drop-down list will consist of the contents of cell B3 to B7. The symbol $ indicates an absolute cell reference.

4. Next click into the Cell link: box and click the cell to the right of the 'user answer' cell created previously. The Combo Box Form will automatically place a number in this cell according to the choice the user has made. For instance, choosing 'ion' from the list will lead to '3' appearing in the 'user answer' cell.

5. Finally enter the number of 'Drop down lines', in this example there are 5.

6. Check that the drop-down list (*ie* Combo Box Form) works correctly and note how the number in the 'user answer' cell changes with the choice. Leave this clicked on the correct answer choice.

The next stage is to use the number output from the Combo Box Form (*ie* the number generated by the user's choice) to generate a response on the User Sheet.

1. In the Admin sheet, enter the number of the correct answer in the cell immediately to the right of the 'correct answer' label. As long as the correct answer has been selected from the drop-down menu on the User Sheet, the 'Difference' cell should now display '0'.

2. Type the alternative messages into cells on the Admin Sheet, *eg*

	D	E
10	Messages	Choose your answer from the list
11		Well done!
12		Sorry, that's wrong so choose again

where the cell containing the text 'Well done!' has relative reference E11 and absolute reference $E$11.

3. Create an IF statement that processes all of the information:

Enter the following without any spaces in an Admin Sheet cell below Q1 and in the same column. The text in quotation marks identifies the cell reference to place in the formula, remember to use the references in the actual spreadsheet created, those given in the example below will not necessarily correspond

=IF('user answer'=1,'choose answer',IF('user difference'=0,'well done','sorry'))

eg =IF(C11=1,$E$10,IF(C12=0,$E$11,$E$12))

The use of absolute ($) references means that if the formula is copied (or 'replicated') elsewhere on the sheet it will still refer to the appropriate cells.

4. Create a cell on the User sheet that displays the feedback created by the IF statement created in the previous step.

Display the User sheet, click in the cell where feedback is required eg below the drop-down menu. Type = in the cell and then navigate to the Admin sheet, click in the Q1 IF formula cell and press enter (ie the return key). The reference should be of the form: ='Admin sheet'!E14

5. Finally test the question and drop-down answer response for Q1.

Creating additional drop-down questions

The two stages for setting up a question and answer (create a combo box then use its output to provide feedback) can be repeated to generate a sequence of different questions. Much of this process can be done by copying the existing Q1 Combo Box Form on the User sheet and replicating the cells for Q1 in the Admin sheet, taking care that the relative cell references are pointing to the correct cells. This has the advantage of maintaining a consistent appearance to the boxes but as absolute references are automatically generated by Excel® when a Combo box Form is created, these references must be re-defined for each new question:

Point at the combo box copied form the previous question, right click the mouse and select Format Control. Enter the

Input range and Cell link relating to the new question, as when the original box was created for Q1.

Creating a 'typed answer' e-worksheet

These use an 'IF' conditional operator. To examine how an existing one of these is set up, eg within the type answer template.xls (resource 32), open the spreadsheet, display the 'answers' sheet and click in cell C13. The formula will be displayed in the Formula bar above the spreadsheet. If this isn't visible, display it by selecting Tools menu, Options, View tab and click in the Formula bar tick box.

The result of this formula is also displayed on the 'worksheet' viewed by students below a white cell. The result for the IF formula changes depending on what text is typed into the white cell.

A limitation of this self-test is that precisely the same text as typed for the pre-set answer must be entered. The only difference allowed is the use of upper or lower case letters. For instance, a space at the beginning of a word would cause the spreadsheet to identify the answer as incorrect.

Use two Excel® sheets within a workbook, designating one the 'User' sheet and the other an 'Admin' sheet.

It is straightforward to edit the template to provide new questions and answers. To create a new one from scratch is more complex. It involves typing an IF formula on the Admin Sheet that provides alternative messages (three in this example) on the User Sheet in response to what the student types into a designated cell. The text in italics in the formula below are cell references:

=IF(*typing*="",*message1*,IF(*typing*=*correct*,*message2*,*message3*))

typing refers to the cell the students type text into on the User Sheet.

typing="" refers to when the student text input cell is blank.

message1 refers to the cell containing instruction text eg 'type answer here'.

correct refers to the cell containing the required text ie different for each question.

message2 refers to the cell containing the approval text eg

'Well done!'.

message3 refers to the cell containing the error text *eg* 'Sorry, the answer is ...'
ie different for each question.

The formula with cell references should appear as follows, the actual cell references obviously dependent on where these have been created:

=IF('user'!D$5="",$C$10,IF('user'!D$5=C6,$C$11,C8))

Creating additional type-text questions

Once the first question and response are created, additional ones may be added by replicating the relevant cells, whilst ensuring cell references are updated. Note that typing, correct and message3 are unique for each question and so require relative references, whereas message1, message2 and message3 are the same and so need to be absolute references.

Modelling changes using spreadsheets

Open and save a new Excel® workbook.

Variable value controls

These can be included using the Forms toolbar.

Select the View menu then Toolbars then Forms to display this toolbar.

Two forms that allow a value to change are paired together, the Scroll Bar and the Spinner. They both work by making a number that can be varied appear in a specified cell.

Click the Scroll Bar button then click and drag the shape somewhere on the spreadsheet. With the mouse cursor pointing at this shape right click and choose Format Control from the menu.

Make sure the Control tab is selected to display the window below.

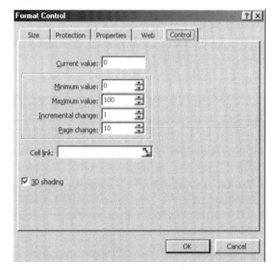

Click in the Cell link box and then click the cell on the spreadsheet where the variable number is required.

The minimum and maximum value for the range of change may be set.

A Spinner form may be set up in the same way as a Scroll bar form.

These controls will only produce integer values. To have a range of decimal values *eg* to obtain an increment of 0.1 use a formula that divides the control value by 10. So if the control value is in cell A1 the value required could be obtained in cell A2 with the formula =A1/10.

Example: a calorimetry model

As an exercise, try creating a calorimetry model for the expression $q = mc\Delta T$

Create Scroll bar or Spinner Forms to give appropriate ranges of values for m and ΔT.

If the cells containing m (in grams) and ΔT (in °C) are B2 and D2 respectively then the formulae =B2*4.18*D2 will calculate q in J.

Graphs

If any values controlled by Spinners or Scroll bars are used to create an Excel® graph or chart then changing the Forms will alter the shape of the graph or chart.

Example: try creating a model for the graph of rate vs [A] for rate $= k[A]^X$

Linking and distributing materials

Giving learners access to materials

There are various ways to do this. Some examples:

- files on floppy disk or CDROM, which is time consuming,

- files on a computer network learners have access to, either in folders or linked to a navigation structure such as that enabled by web pages,

- providing the opportunity for learners to email themselves files

- via a school or college website

- via a 'user group', free space on a web server that allows file and information sharing *eg* **http://uk.groups.yahoo.com/** (June 2004)

Packaging several folders and files together as a single 'zip' file

- the free product ZipCentral **http://zipcentral.iscool.net/** (June 2004), this same software is also needed to unpack the 'zipped' files.

- EnZip from **http://website.lineone.net/~chris_m/ download.html** (June 2004) is a freeware programme for converting zip files into self-extracting (.exe files), which means no software is needed for the 'un-zipping' process.

- WinZip is the original software to create these files, from **http://www.winzip.com** (June 2004) and is relatively inexpensive. It can be upgraded to a version which makes the self-extracting files.

Hyperlinks

Hyperlinks are areas that can be single left-clicked with the mouse to move to a different location. For instance, hyperlinks might be to another part of the same file, open a different file or open a page on a website. Both text and images can be set to act as hyperlinks.

Including hyperlinks within resources can be a very effective way of connecting different materials, or creating a navigable framework to find them within. Ways to create hyperlinks to different files, other points within the same file or websites are described elsewhere within these guidelines for Word® and PowerPoint®.

RS•C

Chemical drawing software (ChemSketch®)

Introduction to ACD / Labs ChemSketch®

This tutorial refers to ACD/ Labs ChemSketch®, free software available from **http://www.acdlabs.com** (June 04). As this can take a long time to download with a slow internet connection, the company have kindly agreed to allow the RSC to include a copy on the CDROM. Drawing molecules using other chemical drawing packages (*eg* CambridgeSoft ChemDraw® or MLD ISIS Draw®) is broadly similar.

After it has been installed, load ChemSketch® by choosing it from the Program menu. Click 'OK' to any windows that appear until the Structure screen is displayed.

ChemSketch® has many features. The content of this tutorial has evolved through responding to the demands of the participants of several RSC sponsored teacher INSET events for teachers of the 11–19 chemistry curriculum.

Before starting it is worth becoming familiar with the few features labeled in the view of the Structure screen below, described over the page. Hover the mouse cursor over a particular button to view a note describing what the button does.

Undo – a most useful feature

Getting used to this kind of software is necessarily an exploratory process which involves unexpected and unwanted things happening. Knowing how to step back if this happens is vital to avoid frustration.

To reverse a change select the Edit menu and click Undo. Repeat this to step back through a sequence of events. Alternatively, use the keyboard shortcut of holding down both the Alt and backspace keys at once.

If a sequence is stepped back too far and it is necessary to move forward again, select Edit menu and click Redo. The keyboard shortcut for this is holding down three keys: Shift, Alt and backspace all at once.

Templates button

'clean structure' and '3D optimisation buttons'

select/move and rotate/resize buttons

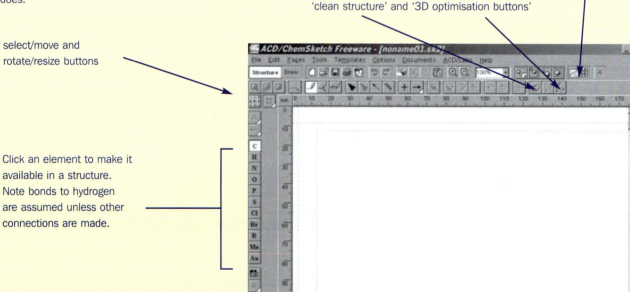

Click an element to make it available in a structure. Note bonds to hydrogen are assumed unless other connections are made.

Drawing 2D formulae

Skeletal and partially displayed formulae

Make sure the C (Carbon) button is clicked. Click once on the screen and CH_4 will appear. Click elsewhere and a second CH_4 appears. Click and drag the mouse from one methane to the other to form a C–C bond and therefore ethane. Click and drag elsewhere to attach another methyl group. With the C button selected, left clicking any C–C bond once converts it to a double bond, clicking again converts a double bond to a triple bond.

Try drawing a hydrocarbon structure containing branches and unsaturated bonds. Note that the default structure is a skeletal formula with only the terminal atoms identified. To tidy up a rough drawing click the 'clean structure' button.

To display all the atoms along the chain follow these steps:

1. Use the Select / Move button or Select All (Ctrl+A) from the Edit menu to select the molecular structure.

2. Open the Properties window by selecting the Tools menu and clicking Structure Properties (or hold down the Ctrl+Shift+S keys all at once). Then select Common and tick All for Show Carbons. The Properties window also allows the colour of individual atoms and bonds to be customised.

3. Click Apply, note this only applies changes to a selected structure. Remember the 'clean structure' button will tidy up any changes.

4. Click the Select / Rotate / Resize button. Click outside a structure and drag the mouse across it to select all of it and then change its orientation and, most usefully, its size.

Clicking a different element button at any time allows the various functional groups to be added.

Changing symbol order in a formula

The default ChemSketch structure may have formulae units written differently from what is preferred eg CH_3 might be preferred as H_3C.

Click the Change Position button, point the mouse at the individual formula unit that is to be changed (eg H_2C) and left click repeatedly until the formula is satisfactory.

eg

CH_3 — HC — CH_3 — H_3C — H_2C might be preferred as CH_3 — CH — CH_3 — CH_3 — CH_2

Practice creating, tidying, customising and deleting a few skeletal and semi-displayed structural formulae until these steps become familiar.

Fully displayed formulae

For example, $(CH_3)_2CHCH_2OH$ might be drawn as:

H H H
H-C-C-C-O
H H   `H
H-C-H
H

1. First ensure all the carbon atoms within the structure will be displayed. Select the Tools menu, Structure Properties, Common tab. Then tick All within the Show Carbons section. Finally click Set Default.

2. With the carbon button selected, and whilst pressing the shift key, draw a propane chain. Holding the Shift key forces the bonds horizontal and only allows fixed lengths to be drawn. H_3C——CH_2–CH_3

3. Still pressing the shift key, draw a methyl side group down from the central carbon atom but drag it two bond lengths clear of the chain. This additional length is essential to prevent side chain hydrogen atoms overlapping with the main chain ones.

   H_3C——CH—CH_3
   |
   CH_3

4. Select the oxygen button and, whilst pressing the shift key, add an OH group to the end of the chain.

$$H_3C — CH — CH_2-OH$$
$$|$$
$$CH_3$$

5. Select the hydrogen button and, whilst pressing the shift key, drag -H bonds from the carbon atoms . Note that whilst the shift key is pressed, only fixed bond lengths may be drawn which maintains a consistent structure.

To fully display the OH bond, select the H button and drag a bond. Pressing the shift key whilst doing this keeps the bond horizontal or vertical. Alternatively, the bond can be set at an angle.

6. Control the length, thickness and colour of the bonds and the font, size and colour of the atoms by selecting the whole structure and opening Structure Properties in the Tools menu.

7. To illustrate the shape around a double bond, first draw an ethene structure, click the 3D Optimization button and then the Clean button

To extend the hydrocarbon chain, convert one of the H atoms to a C by pointing the mouse at it and clicking. Then use the techniques described previously (1 to 4 above) to draw and display the rest of the structure.

Structural formulae

Formulae such as $CH_3CHClCH_3$ can be written using ChemSketch:

1. Switch to the Draw view screen (button towards the top left of the screen).

2. Select the Text button towards the bottom left of the screen (remember: hovering the mouse over a button identifies it). Click and drag a text box on to the screen.

3. Type a formula, making use of the Subscript (S-) and Superscript (S+) buttons immediately above the screen.

The completed structural formula is a ChemSketch picture which can be copied and pasted into another application by using Paste Special, unless a link to load ChemSketch is required.

Charges and free radicals

To add charge to a structure or indicate a free radical select the atom to be given charge and then click the Increment (+) Charge or Decrement (-) Charge button:

Clicking the bottom right triangle of this allows negative charge, unpaired electrons, or combinations of these to be displayed:

$$H_3C — \overset{+}{C} = O \qquad H_3C^\bullet$$

Aromaticity

For example, draw a benzene 'Kekule' structure:

To change this to a representation of the delocalised ring, first select the structure then select Tools menu, Show Aromaticity to give:

which can be converted to:
using Tools menu, Structure Properties, Common tab and choosing not to 'show carbons'

ChemSketch includes several templates, and one of these is for a range of pre-drawn aromatic molecules. Select the Templates Window from the Template menu (or press the F5 key or the Open Templates Window button). Select the Aromatics template from the drop-down menu.

Functional groups and hydrocarbon chains may now be added as required:

Side group/substituent formulae without bonds

To display a side group without bonds, eg to display the carbonyl substituent in the previous example as $COCH_3$:

1. Draw a benzene ring with a methyl group substituted for H:

2. Click the Edit Atom Label button ('abc' below the element buttons), then click the methyl group to reveal the Edit Label menu.

3. Either choose a label from the list or type a new one, eg type $COCH_3$.

4. Click insert. Any subscripts should be recognised as such:

Drawing enantiomers

To create an enantiomer from a structure drawn using the 3D 'stereo' wedge bonds:

1. Draw an enantiomer illustrating the chiral centre using the up and down 'stereo' representations as well as the normal line representations.

2. Copy and paste a structure drawn using the wedge bonds

3. Switch off the 'Keep Stereo configuration on Flips' in the Options menu, Preferences, Structure tab.

4. Select the copy and press the 'Flip Left to Right' button to create its enantiomer.

The resulting structure is the enantiomer of the initial structure.

eg 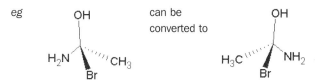 can be converted to

Naming structures

ChemSketch includes a feature that automatically generates a name for a selected structure.

Select any structure and click the Generate Name button:
If there is only one structure on the page there is no need to select it , simply click the button.

Using templates, including laboratory apparatus

To the extreme right of screen is the Table of Radicals button. Click this (or press the F6 key) to display its window. This provides a wide range of pre-drawn structures or useful fragments to build from.

There are a large number of other templates available from the Template Window. Open this in the Templates menu (or by clicking its button or pressing F5) and selecting a choice from the Template List drop-down menu.

RS•C

Two useful examples are:

• the seven pages of apparatus diagrams in the 'Lab Kit' template examples:

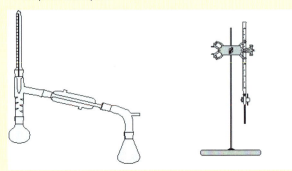

• lone pair electrons in the 'Lewis Structures' template,

simply select one of these and then click where these should appear in the structure.
eg added to a structural formula to create:

$$CH_3\overset{..}{N}H_2$$

Saving ChemSketch® screens

To save a ChemSketch® screen, choose the File menu and then Save As. Navigate to the folder where you want this to be stored by double clicking on the visible folders until the one required appears in the Save In box.

If you want to create a new folder, navigate so that the folder to contain the new folder appears in the Save In box, click the Create New Folder button and name this. Open the new folder and save the ChemSketch® file as above.

The screen is saved as a ChemSketch® file with 'sk2' as the file extension.

Pasting drawings into other applications

Once a drawing is completed it can be saved or simply pasted into the application where it is to be used eg a Word® document or PowerPoint® slide.

There are two ways of pasting into another application:

1. As a ChemSketch object. Pointing the mouse at a ChemSketch object and double clicking launches ChemSketch, if it is available, so that the drawing can be changed. To do this simply copy (select the drawing in ChemSketch and press the Ctrl+C keys together) and paste (open the application then press the Ctrl+V keys together). If this option is used there is no need to save the drawing as a ChemSketch file.

2. As a picture. Copy as above (Ctrl+C) then switch to the application where the drawing is required, open the Edit menu, select Paste Special and choose Picture. This then appears as a simple image.

3. To position the picture on the page, point at it with the mouse and left click to display a menu. Select Format Picture, Layout and eg In front of text.

Review exercises

1. Draw a structure for 1-amino-2-methylpropane. Display all of its atoms, tidy the structure and add a lone pair of electrons to the nitrogen atom. Select the whole structure and make it larger. Copy it and paste to another application.

2. Assemble the apparatus for a reflux experiment, copy this and paste into another application.

Converting to 3D interactive molecular structures

ACD/ChemSketch includes a 3D Viewer that produces a range of images of structures drawn in the 2D mode. It does not represent non-bonding electrons.

1. Draw a molecule on the Structure screen.

2. Select the ACD/Labs menu and click 3D Viewer. This loads the viewer and displays the structure as 'balls and stick' with hidden H atoms. Click the 3D Optimisation button to reveal the H atoms. Remember that hovering the mouse over a button displays a text description.

3. Explore the various views by clicking the buttons above the screen or simply right clicking the mouse. Note the 'With Dots' option allows the various representations of atoms and bonds with a (very) rough representation of the extent of electrons superimposed.

4. Note that the molecule can be rotated by dragging it with the mouse. Clicking the Auto Rotate makes this happen automatically. Clicking the Auto Rotate and Change Style button causes the molecule to loop through the various representations whilst rotating, which is a useful screen for an Open Evening.

5. To change the default colours of the atoms or background click the Set Colours button. Although the range of colours available is very limited it can be useful to set the atom colour to match, for example, the Molymod model colours so that images on the screen look like the models students are handling in the classroom.

Pasting 3D structural images into other applications

3D images can be pasted into other applications as still pictures *ie* with no interactivity. As above, there are two ways of pasting an image into another application. Copy and paste (Ctrl+V) produces ChemSketch objects which, as long as ChemSketch is available on the computer, may be double clicked to launch ChemSketch and then edited. Copy and Paste Special (available in the Edit menu) allows the image to be pasted as a Picture, which cannot then be edited in ChemSketch.

The 2D ChemSketch structure on the left has been converted to the 3D ball and stick with dots image on the right. Note that the 3D image has been given a white background in ChemSketch.

Review exercise: Try creating 2D and 3D images of propan-2-ol and paste both into another application.

Adding interactive 3D structural images to web pages

The previous section dealt with pasting 3D images into other applications as still pictures which meant that interactivity was lost. It is possible to export structural drawings as mol files which can be incorporated within web pages as interactive images as long as suitable 'plug-in' software has been loaded. MDL Chime® is an example of a plug-in that is widely used by chemists for viewing and manipulating 3D structures. It is available to download free from the Chime website at: **http://www.mdlchime.com/** (June 04)

One advantage of creating Chime images over the ChemSketch 3D Viewer images is that Chime® supports a representation of multiple bonds.

To create a 'mol' structure image file using ChemSketch:

1. Draw a structure in ChemSketch and click the 3D Optimise button whilst still using the Structure screen.

2. Go to File menu, Export, then select the location where the mol file is to be saved. In 'Save as Type' choose MDL Molfiles (*.mol). Type in a suitable filename and click on Save.

3. The file is now usable in Chime.

To view the structure in a browser, provided the Chime plug-in is installed on the computer being used:

1. Create and save a mol file.

2. Open the folder where the mol file has been saved and double click its icon. Ideally, a web page browser (eg MS Explorer®) will launch and then the structure will be seen, possibly only in the 2D view. If instead of a browser opening, ChemSketch simply re-loads, then try opening the browser and make the folder containing the mol file visible at the same time. Use the mouse to drag and drop the mol file on to the browser window and the file should now open in the browser.

3. Point the mouse cursor at the structure and right click to display a menu which allows the structure to be manipulated:

   If the molecule is displayed as 2D click on 3D Rendering. Display obtains different views.

Rotate causes the displayed structure to rotate.

4. Holding the shift key whilst mouse dragging the structure increases its size.

To add the CHIME structure to a web page use web authoring software. eg with Microsoft Front Page®:

1. Open an existing web page or create a new one.

2. Go to the Insert menu and select Web Components. In the Component type window select Advanced Controls (NB with versions of Front Page before XP select Advanced from the Insert menu).

3. In the Choose a control window select Plug-In.

4. In the Plug-In properties window browse to the mol file to be inserted, set the height and width to display the molecule as required and click OK.

Web links

This section is available from the ChemIT website or the CDROM (resource 99). Clicking links within this file is much quicker than typing them out.

Remember copyright considerations when making use of these sites.

http://www.ChemIT.co.uk is the website set up to support the RSC Teacher Fellow ICT project. Download the materials and use them with students. Contribute new materials to help build a shared resource for all chemistry teachers.
All links tested June 04.

Internet search engines:

http://www.google.co.uk for specific searches and finding images

http://www.kartoo.com for general topics or vague searches

Matisse Enzer's Glossary of Internet Terms:
http://www.matisse.net/files/glossary.html

Oxford University's 'Screensaver Lifesaver':
http://www.chem.ox.ac.uk/curecancer.html

RSC supported:

1. The RSC has its own education resources website, LearnNet:
http://www.chemsoc.org/networks/learnnet/index.htm
where the project material will also be available at
http://www.chemsoc.org/learnnet/ChemIT.

2. Earth Sciences – the JESEI (Joint Earth Science Education Initiative) project.
http://www.chemsoc.org/networks/learnnet/jesei/index.htm

3. Green Chemistry: This is a post-16 resource to build on the book 'Green Chemistry' by RSC Teacher Fellow Dorothy Warren sent to all schools in 2001:
http://www.chemsoc.org/networks/learnnet/green-chem.htm

4. Chemistry and Key Skills: Exemplars and guidelines illustrating how chemistry practical work can be used as a source for Level 3 coursework for IT and Number: **http://www.chemsoc.org/networks/learnnet/keyskills.htm**

European Chemical Industry Council website: **http://www.cefic.org**

For a very wide range of Chemistry resource downloads developed by teachers go to **http://ferl.ngfl.gov.uk/** then follow the links:
learning & teaching / browse by subject / sciences / chemistry

It is also worth looking at the 'content creation' area.

A set of revision quizzes mapped to each of the three main A-level specifications: **http://www.mp-docker.demon.co.uk/** and video clip site **http://www.mpdocker.demon.co.uk/**

A database of copyright free spectra: **http://www.aist.go.jp/RIODB/SDBS/menu-e.html**

A database created by UK Chemistry teachers as a source of drawings and diagrams **http://www.btinternet.com/~chemistry.diagrams/**

Chemists' Net is a good source of resources and contacts, including an e-mail discussion forum for Chemistry teachers: **http://members.lycos.co.uk/chemistry/**

The Schoolscience site calls itself 'the leading site for free content about school science and its applications': **http://www.schoolscience.co.uk**

The University of Oxford has produced a series of virtual Chemistry experiments using both streaming video and animation. It has an on-line post-16 textbook and some very swish virtual tours of Oxford : **http://www.chem.ox.ac.uk/vrchemistry**

Exemplarchem: An internet based competition of exemplary project work in chemistry related website design for UK undergraduates.
http://www.chemsoc.org/exemplarchem/index.htm from where entries can be accessed eg the Flash 'alchemy' tutorial linked below.

The Reciprocal Net is a crystallographers database. It includes contextual information about each substance. Needs Sun's Java plug-in, vs 1.2.1 or later, to see the complete structures: **http://www.reciprocalnet.org/common/index.html**

Salters' A level Chemistry site: **http://www.avogadro.co.uk/**

Nuffield A level Chemistry site: **http://www.chemistry-react.org**

Macromedia Shockwave, Flash and Director:

The University of Oxford on-line post-16 text book uses Flash throughout :
http://www.chem.ox.ac.uk/vrchemistry/foundation.html
eg states of matter:
http://www.chem.ox.ac.uk/vrchemistry/theview/html/page07G.html
Flash 'alchemy' tutorial:
http://www.chemsoc.org/exemplarchem/entries/2002/crabb/flash.html

Chemland (© 1999 Harcourt Brace & Company. All rights reserved) has a large number of Flash animations:
http://www.harcourtcollege.com/chem/general/common/ice/chemsim.htm
eg it has an equation balancer at:
http://www.harcourtcollege.com/chem/general/common/ice/chemsim/balancing_win.htm
Lots of material but of variable quality (Thermochemistry: NaCl in water is excellent) at:
http://www.chem.iastate.edu/group/Greenbowe/sections/projectfolder/animationsindex.htm

Laboratory Simulation Software

IrYdium Project VLab is free
http://ir.chem.cmu.edu/irproject/applets/virtuallab/
but it is necessary to a download a very large (16 Mb) zip file.
Two commercial products are: ChemLab from
http://modelscience.com/ and Crocodile Clips from
http://www.crocodile-clips.com/chem.htm

RS•C

Sources of material to enliven or inspire, some less than serious:

WebElements Periodic Table:
http://www.shef.ac.uk/~chem/web-elements/

molecule of the month:
http://www.bristol.ac.uk/Depts/Chemistry/MOTM/motm.htm

molecules with silly names:
http://www.bristol.ac.uk/Depts/Chemistry/MOTM/silly/sillymols.htm

Theo Gray's brilliantly eccentric Periodic Table site:
http://www.theodoregray.com/PeriodicTable/index.html

The ChemIT website

The website at **http://www.ChemIT.co.uk** is currently (June 2004) being maintained as the source of ICT based materials and ideas generated through the 2002–03 RSC Teacher Fellow ICT project to support chemistry teaching.

It will be supported so long as teachers continue to make use of its contents and supplement these with new ones of their own. The resource bank already includes many contributions from different teachers, and it will develop to become ever more useful as a shared resource if this cooperative effort continues.

If the Ideas and Skills guidelines published as a result of this Teacher Fellow project lead you to produce any materials or approaches, and you find that these work well with your students, please feel free to contribute these by contacting the website. This may be for the entirely pragmatic reason that you will have access to new contributions from others.

The chief criterion for the inclusion of a new resource is its chemistry content, and certainly not the level of technical sophistication used to create it or, within reason, the quality of presentation.

These 'Skills' guidelines should be seen as a current view or 'snap shot' of how the software most widely available is being used by teachers of chemistry, based on feedback from a significant number of colleagues. The ideas and approaches are constantly being refined and supplemented.

For instance, the suggested use of Lookup Tables in Excel® as a straightforward and elegant way to provide feedback to students developed from the activities (also described within the uses of Excel section) that make use of drop-down (combo box)_ and radio button forms to provide interactive choices for students. It was only added to these notes immediately prior to publication, and is an illustration of how approaches can develop through trying things out with students and exchanging ideas between teachers.

http://www.ChemIT/co.uk provides an opportunity to share new ideas and approaches, particularly as inevitably some of the skills and suggested activities set out in this document will be superseded.